This book is for performing a method of self-healing and for the development of ones own consciousness. This method can only be used in appropriate self-care. The book offers no alternative for professional treatement of severe illnesses and it should not stop you from taking necessary investigations by conventional medicine.

ABOUT THE BOOK:
Can vitality and spirituality of a person be demonstrated by physical measurements? This question has preoccupied Michael König for over 30 years. During his student days, he built his first Kirlian device; in the course of his professional life as a physicist, he carried out studies on the human energy system. In doing so, he discovered that the electrical charge of a persons cells is closely related to his vitality and state of consciousness. König developed this modern diagnostic method over the decades, which can be of assistance to physicians in order to better understand the health status of their patients: the photon diagnosis.

ABOUT THE AUTHOR:
Dr. Michael König, born in 1957, is quantum physicist and has dedicated nearly 30 years to the exploration of the relationship between mind and matter. From 1987 to 2004 he directed a private research institute and obtained patents in the field of complementary medicine. As one of the pioneers of the new physics and the shift of paradigm he is a frequent speaker and lecturer at international conferences, at universities and in documentaries.

© Michael König

Dr. Michael König

A Photon-Diagnosis

Vitality is measurable – how alive are you really?

Bibliographical Information of the Deutsche Nationalbibliothek
This publication is listed in the Deutsche Nationalbibliographie of the
Deutsche Nationalbibliothek; detailed bibliographical information
can be accessed under http: //dnb.d-nb.de

© 2014 Michael König
Printing and Production: BoD – Books on Demand
ISBN: 978-3-7357-1090-1

Contents

Introduction   7

CHAPTER I
The discovery of electroluminescence
of living objects   9

CHAPTER II
Applications of the Kirlian effect in the West   19

CHAPTER III
First personal experiences
with the Kirlian effect   32

CHAPTER IV
Without biophotons there is no bioplasma   50

CHAPTER V
The Kirlian effect on the way
into the digital age   60

CHAPTER VI
The photon diagnosis   69

CHAPTER VII
Results of initial case studies   90

Summary   114

Literature   116

Introduction

Since the discovery of the kirlian effect, in the former Soviet Union, Kirlian photographs of animate and inanimate objects fascinate the viewer, the specialist as well as the scientific layman.

Since my studies of physics began over thirty years ago, I started to explore the physical characteristics of biological processes.

Initially, it was my hypothesis that electrical charge and the electric and magnetic fields caused by them are associated with both our physical and our psychological state.

In this book I will also report on first applications of Kirlian photography in the Soviet Union in the fields of geologicy of mineral deposits, agriculture and medicine as well as on the further development of the process in the West since the 1970s.

Modern quantum physics and high-tech electronic equipment make this possible: With the fascinating process of photon diagnosis vitality, psychological state and also the personal spiritual transformation level of patients based on their electromagnetic state are now objectively measurable.

In Chapter VII the first results of ongoing studies in medical practices are presented that have already been applied from my the Kirlian photogaphy furthering the development of digital diagnostic methods in several patients.

CHAPTER I

The discovery of the electro-luminescence of living objects

From its beginnings into the modern era, whilst in the dark, sailors observed a strange bluish glow that most vigorously appeared at the top of the mast of their sailboat. Usually, on the mainland such observations were made as well, mainly at church spiers or from other vantage points. Most of these luminous phenomena were made before or during a thunderstorm at night, so it was natural to associate them with processes in the atmosphere.

Later, the phenomenon was commonly known as St. Elmo's fire and with the scientific investigation and a more comprehensive description of electricity, one could find a natural and plausible explanation for these luminous phenomena. From the physicist and naturalist Georg Christoph Lichtenberg, who lived and taught in Göttingen in the 18th century, original records are still available: According to him, the discharge channels, as they occur in the Kirlian Photography as well, are referred to as Lichtenberg figures. In physics, these processes are summarized under the term of electro luminescence; nowadays, the process whithin the parameters of quantum physics are well understood. Today we know that these processes occur in nature whenever there is a build up of a large electrical charge difference in the atmosphere. Before such differences are discharged in the form of a lightning bolt, smaller discharge currents are flowing, especially on to sharp objects because there the electric potential differences are the greatest.

With the ever growing use of electricity by people in everyday life, it was only a matter of time before the electroluminescece would also occur in technical devices where high electrical voltages are used. The early Soviet-Union undertook particularely ambitious projects in order to advance the electrification of the Communist Society. Lenin expressed his political opinion in the simple formula: Socialism is equal to Soviet power plus electrification. And so it is perhaps no accident of history that it was precisely a soviet citizen who should succeed in discovering the electro luminescence in living objects. So it was the electrical engineer Semyon Kirlian who, in the 1930s, when repairing an electrical apparatus in which the internal high voltage was poorly insulated in relation to the outer housing, made the observation that from his hands emenated a strange bluish luminescence when he touched the defective machine which was under electrical power. Soon after this, he started to investigate this phenomenon together with his wife Walentina in detail. They found that other living organisms, including animals and plants, radiated these strange luminous phenomena as soon as they came into contact with a high voltage and high frequency field.

Shortly after this, soviet scientists became aware of the discovery Semjon Kirlians and others began to explore the effects using scientific methods. The totalitarian system under the rule of Stalin made it very quickly a state secret and so the knowledge about it came into the West only in the 1970s. Not much is known about the Kirlian research in the Soviet Union before the Second World War, but it is well known today that the Kirlian effect was used there already in the 1950s for medical diagnosis in hospitals, for seed and harvest control in agriculture, for the locating of mineral deposits in geophysics and also in the material sciences and material testing.

Soviet physicians and life scientists recognized early on the potential of Kirlian photography in medical diagnosis. They coined the concept of bioplasma in medical diagnostics and thus recognized that the Kirlian photograph of a patient is caused by electrical charge interactions inside the body of a person. The intensity of the radiation corona which occurs in the Kirlian effect depends of course directly on the conductivity, i.e. the amount of free or mobilizable electrical charge carriers. The higher the amount of charge carriers present, the lower the electrical resistance and the higher is the electrical conductivity and vice versa. The scientists realized that the Kirlian photography is ideally

11

suited for therapy monitoring, by taking a Kirlian picture before and after a treatment and then comparing the two. However, this purely quantitative comparison of two radiations is not the only criterion for judging such Kirlian photographs, because you have to consider the radiation quality, i.e. the characteristics of the rays of the corona.

In the agricultural sector of the Soviet Union seeds were studied for their suitability in different climatic areas. Namely the emission of a seedling could be selected depending on how well it is suited for a particular growing zone or whether other seed varieties would have better growth prospects.

During the exploration of ore deposits was first searched for broad hints in changes in the earths magnetic field when flying over an area. Then, one took soil samples at points of interest and examined them with the Kirlian photography in context to their radiation behavior. Depending on the mineral composition the study of each sample showed detailed reference to the ores contained.

In the material sciences the emission of a material sample was examined with the Kirlian photography or one could also detect the presence of hairline cracks in individual works.

In the following we will look at the physical processes in the emergence of a Kirlian photograph in a little more detail. Accordingly, we must first take a look at the interior of a classic Kirlian device. Such appliances typically consist of an electrically insulating plastic housing, in which an electronic circuit generates a high frequency pulsed high voltage. This is connected to a sheet-like metal electrode, which is arranged below the electrically insulating cover plate of the device.

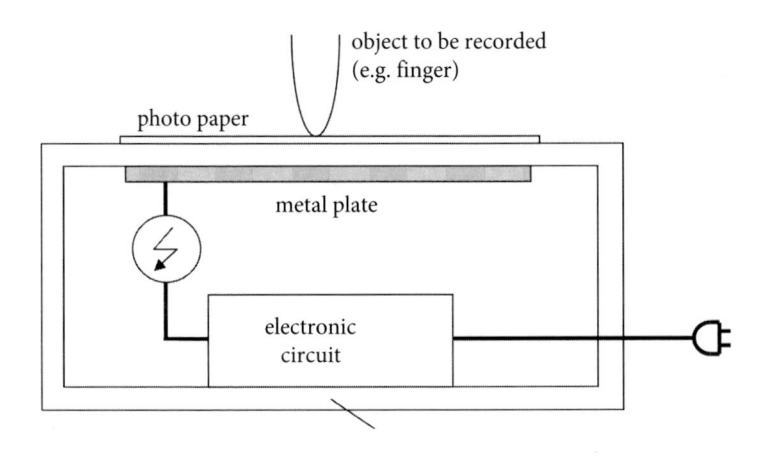

Figure 1 Cross section of a classical Kirlian device

The insulation guarantees that one does not come into direct contact with the dangerous high voltage which is touching the metal electrode during the operation of the device. The cover plate has the property of transmitting the electric field generated by the high voltage in spite of the insulating. Now, if you put a photo paper between the object to be photographed, such as a finger, and the cover plate, in the dark you can map the highly visible bluish radiations of the finger on the photo paper and thereby expose it photographically. Then you have to develop the photo paper in a darkroom process, dry it before it is ready as a Kirlian picture.

How do the luminous phenomena occur when turning on the Kirlian device? This process, which is referred to in physics as electroluminescence generally or specifically as a gas discharge, can be explained by quantum effects. As soon as the pulsed high voltage is turned on at the Kirlian device, a uniform electric field builds up on the top of the plate. If an electrically positive voltage is applied on the metal electrode in this moment, electrically negatively charged particles, which are contained in the object to be photographed, become mobilized by the electric field and accelerated towards the metal electrode. Because oppositely poled charges (positive – negative) attract each other. If there is an electrically negative voltage on the metal electrode at this moment, so positively charged particles will be accelerated towards the metal electrode out of the recording object. On their way through the air, mobilized particles constantly collide with air molecules.

At each impact with an air molecule an electrically charged particle will give a portion of its kinetic energy that was received in the electric field, to the air molecule and thereafter is again accelerated in the electric field until it collides again with another air molecule. This process is repeated for each charged particle from several hundred to a thousand times, until it finally dashes against the photo paper and remains inserted therein or slowly migrates through the electrically insulating cover plate. When an air molecule now absorbs energy of a charged particle in a collision process, an electron bound in the molecule leaps to a higher orbital for a short time.

It mostly takes only a few millionths of a second in which an electron remains in the higher energy orbital. During this tiny period of time the air molecule is in a so called excited state. Then the electron in the excited molecule tumbles back to its original orbital and is again in its original state. In doing so, according to the law of conservation of energy, it has to re-emit the energy abosrbed from the colliding particle. If the energy received in such a collision process was not high enough to remove the electron from the molecule, the electron in return will emit the absorbed energy in the form of electromagnetic radiation. In other words, a physicist would say that the electron emits a photon when it returns to the ground state. Photons are quantums - they are the quantums of electromagnetic radiation. The energy of the photon is thereby typically as big as it corresponds to the energy of visible light. The photon is in this case a light particle and therefore in this manner the photographic paper is exposed and in dark rooms we can observe this process well with the naked eye.

It follows that a Kirlian picture primarily contains information on the number of electrically charged particles that are located on or near the surface of the recorded object and how they are distributed. Figure 2 illustrates this state of affairs.

In **1** shows an excited atom or molecule in the interior of the recorded object, such as a finger of a patient. In an excited molecule only a small portion of energy is needed to mobilize an electron from that molecule by means of the strong electric field. This quantum physical process is called field emission. In figure 2 the above in detail explained quantum-physical situation of shock excitation, followed by emission of a photon is described, which results in the exposure of the photo paper. Figure 3 shows the situation in an atom or molecule in the interior of the recorded object which is in the ground state. Here, even the presence of the strong eletric field on the cover plate of the Kirlian device is not sufficient to mobilize an electron.

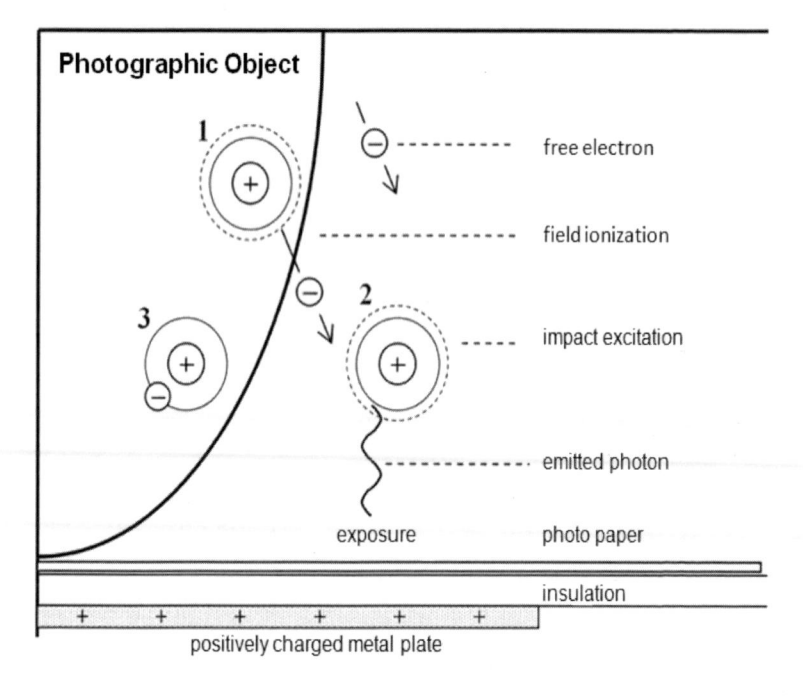

Figure 2 The path of charged particles on their way through the air to the cover plate of a Kirlian device

To better understand the structure of the recorded radiations in a Kirlian picture and to connect them with the realities in the captured object, further biophysical and quantum mechanical skills are required which will be further explained in the following chapters of this book.

During the electrical discharges that occur with the Kirlian effect, not only excited, but also partially ionized molecules arise to a certain percentage. Ionized means that an atom or molecule has lost one or more electrons (electrically positively charged ions), or that the atom or molecule has one or more surplus electrons (electrically negatively charged ions).

If many atoms or molecules of a gas are ionized, it is called a plasma. Plasma is often referred to as the fourth state of aggregation, in contrast to the solid, liquid and gaseous state of a substance.

This is because in a plasma many more electrically charged ions are present than in a gas, much more electrical charge can be transported in the plasma, which is nothing more than an electrical current.

However, if a Kirlian photography is made, typically only a very small current flows, such as a few millionths of an ampere, and then only during the short exposure time which is a few seconds usually. These are quantities of charge as they occur for example, when we take off or put on a sweater which is electrostatically charged and thereby hear the crackling or even see the individual micro discharges in the dark. We know the phenomenon also from electrostatically charged synthetic carpets. When we walk on such a carpet and then touch a doorknob, an electrostatic discharge happens in a split second. This is umpteenth-fold more intense than the micro-discharges during a Kirlian photography. By normal operation of a Kirlian device there is, therefore, as with other electrical machines, no risk of getting a fatal electric shock.

I have not come across a single case where a person got a dangerous electric shock while taking a Kirlian picture in the past 30 years. Nevertheless, I always advised all equipment vendors not to take pictures with a Kirlian device of individuals with electrical implants, particularely pacemakers.

CHAPTER II

Applications of the Kirlian Effect in the West and the Energetical Terminal Point Diagnosis (ETD)

Finally, in the 1970s a part of the knowledge of the Kirlian effect that had been collected in the Soviet Union since the 1930s came through the Iron Curtain into the West. At least the necessary details were known how to create the luminous phenomena that occurred during the Kirlian effect by a suitable electronic circuit. Then in 1978 the Kirlian photography was officially declared a state secret in the former UdSSR. From the US it became known that there were also scientists that began to deal with the Kirlian photography; in particular the space agency NASA was said to have created a seperate department specifically for this purpose.

From now on, esoterically interested circles plunged themselves particularly eagerly into this technology. Thus it was hoped to find through it an explanation for the human aura. And so, up to this day, the statement is still rumored by many people that with the Kirlian photography one can make visible the human aura. Partially the process was then offered as aura photography, which has not done much good to the assessment of the scientific validity of Kirlian photography from the beginning. So-called phantom images, in which cut leaf tips still remained visible in the Kirlian picture augmented the myth that with kirlian photography an access had been found to the supernatural. However, in initial studies in the early 1980s, I came to the conclusion that at best it was a matter of sloppy specimen preparation, possibly even a deliberate attempt to deceive. In a later chapter I will go into this in more detail.

If we do not define the human aura as something supernatural, but as the sum of the electromagnetic radiation of man, we get closer to the matter. As I have explained in the last chapter, the more electric charge carriers are emerging from a biological object, the more atoms or molecules of the object are already in excited states. And when atoms return to the ground state from their excited state, the excess energy in the form of electromagnetic radiation, ie photons, is released. In this respect soon after the announcement of the Kirlian photography a connection with the biophysics arose also very quickly and in particular with the biophoton research which had just started in the 1980s. A doctoral student of the german physicist and professor at that time at the University of Marburg, Fritz Albert Popp, was one of the first to have succeeded in proving the evidence of the existence of the ultra-weak cell radiation by making highly sensitive measurements on cucumber seeds. The goal of this research project was originally to falsify the existence of this radiation, ie to show that it just did not exist. Instead, this radiation turned out not to be the white noise of a chemoluminescence which is the waste product of chemical reactions, but various biophysicists led the experimental proof at that time, that this radiation is even coherent, ie of a high degree of order and probably serves the electromagnetic control of the metabolic functions of cell components.

These findings about the coherent nature of the ultra-weak cell radiation provided a strong argument for the relevance of the measurable radiations occuring during Kirlian photography and its relationship with metabolic processes in the organism under examination. The biophysicists around Fritz Albert Popp and their collegues from abroad proved by evidence in numerous scientific publications that pathogenic, ie abnormal changes in the cell go along with a disturbance or blockage of the discovered ultra-weak cell radiation.

The famous physician Rudolf Virchow who co-founded the modern pathology in the second half of the 19th century recognized already more than a hundred years ago: "The nature of illness is the pathogenic modified cell".

Here lies the greatest potential in the use of Kirlian photography, namely in its application in medicine. The energetic and not purely material view of the nature of man was unfortunately more and more lost in the course of the development of medicine.

We can only hope – and it is also to be expected from a bio technological and quantum physical point of view – that this trend will be reversed in the future, for it has long been clear that through the surgical interventions and the administration of chemical substances the level of the very health of man is not to be raised.

Only a few users of Kirlian photography have obtained a certain degree of fame since the 1980's and some of them have deservedly rendered outstanding further development of Kirlian photography in the fields of technology and application. This includes the graduated physicist Dieter Knapp, who among other things had a metal vapoured transparent high-voltage electrode patented and developed his invention, the plasma-print method, with which he created coloured Kirlian shots. This method became particularly well known through books on homeopathy.

A working group of the Department of Physics at the Technical University of Berlin examined in the late 1980s until the early 1990s with a homemade device the reproducibility of Kirlian shots and the influence of various parameters such as the frequency of the high voltage and the exposure time. Mainly pictures of plants, fingers or metal objects were produced. In their documentation the working group came to the conclusion that a good reproducibility could be achieved with recordings of fingers. So picture structures typical for a person could be identified and also influenced specifically over months.

In particular, the working group was able to

determine that different pressure during the touching of a finger on the electrode and a variation of the tilt angle (45-60°) do not substantially alter the image structures. When parameters were maintained consistantly, images of fingers could be reproduced quite well. The influences of the essential simple parameters (touch down angle and touch down contact pressure, frequency and exposure time) for the Kirlian photography of human fingers could be investigated and controled and so, standardized recording conditions were chosen that ensure a high reproducibility.

In complementary medicine the medical practitioner Peter Mandel became known for his use of Kirlian photography in the energetic diagnosis. The Energetical Terminal Point Diagnosis (ETD), which he developed can be considered as a diagnostic standard method due to the wide spread use and acceptance of it, which is based on the Kirlian photography. Therefore, I would like to enter into the details of this process more comprehensively.

In the ETD all the fingers of both hands and after that all the toes of both feet are recorded simultaniously with a Kirlian device on a photo positive paper of the size a little bigger than DIN A4. The exposure time amounts to about 3 seconds for the fingers and 6 seconds for the toes. The recording and subsequent development happens just like in any classical recording method under darkroom conditions. The evaluation of the recording according to the rules of the ETD can be done only by a medical practitioner or physician familiar with and trained for this procedure. Therefore, the ETD is usually applied in a medical practice. The method relies on empirical experiences of acupuncture teachings known from Traditional Chinese Medicine (TCM). This is based on the assumption that the human body is crossed by several so-called meridians, the ending points of which are located at the finger and toe tips respectively. Therefore, these are referred to as terminal points. However, there are suposedly also links existing between the meridians. The function of the meridians is one of energy channels that distribute the so-called Chi, a kind of life energy, in the organs of the body. Chi occurs in the form of two polarities that are referred to as Yin and Yang.

Against this background a Kirlian photograph as described above should give further information about the energy distribution of the chi energy according to the ETD. Not only the distribution of the energy is taken into account based on the intensity of the radiations at the different terminal points of the ETD, but also the quality of the radiations themselves. The ETD makes hereby no statements about the physical nature of the radiation but provides a purely phenomenological description. In the ETD basically three mutually definable radiation qualities are distinguished, whereby entirely smooth transitions between them exist.

The first radiation quality is referred to as quality-normal or endocrine radiation. It can be recognized in the Kirlian picture as fine, tuft-like discharge channels, pointing radially away from the respective terminal point. With this type of radiation, the more harmonious it is distributed around a terminal point, the more it is associated with a healthy state of the organism.

The second radiation quality is defined as toxic radiation quality. In the Kirlian picture you can see it as point-like condensations of the radiations and it will be given even more attention in the evaluation, the further these condensations are moved outward away from the inner aureole of a terminal point. This radiation quality occurs particularly with inflammatory processes. This can be a local focus point happening, if in a particular organ or in a certain body zone an inflammatory process is in progress, or if in the entire body toxic substances are circulating through the body fluids, ie in lymph or blood. In the latter case, this radiation quality is distributed across all sectors in almost all terminal points. Toxic radiation quality can also be caused by a viral load, which is a result of an infection.

Poisoning through all types of chemicals can also be shown in the Kirlian photography by the toxic radiation quality. How well certain strains can be specifically recognized however, is controversial among the users of the ETD.

The third quality of radiation is referred to as degenerative radiation quality. Appearance wise it is an often powerful, but completely structureless emission. It is interpreted in the ETD as energetic rigidity and

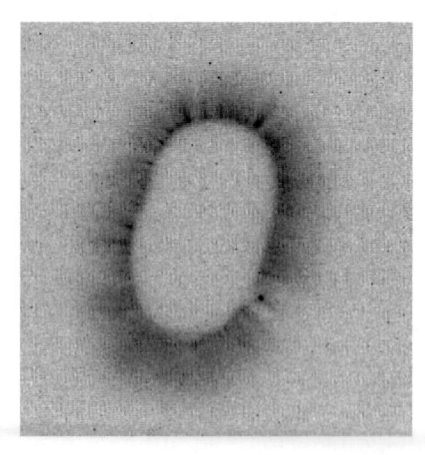

Figure 3: The three radiation qualities in the Kirlian picture according to the Energetical Terminal Point Diagnosis (ETD) in the example of a finger

a) Normal radiation or endocrine radiation quality

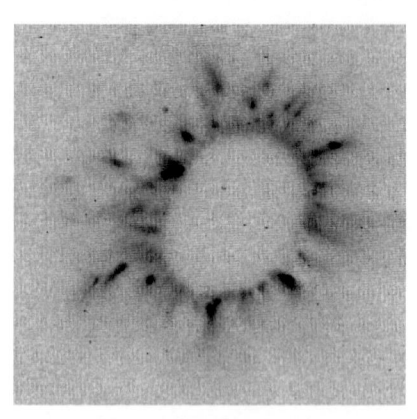

b) Toxic radiation quality

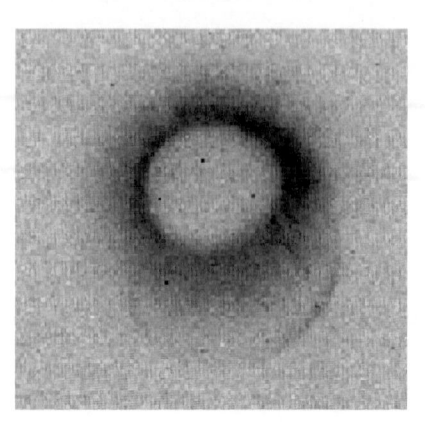

c) Degenerative radiation quality

regulation disability and is often an indication of severe metabolic disorders or diseases for users. It occurs particularly often if the body is strongly over-acidic and flooded by waste materials of metabolism.

The ETD is not a therapy in itself, but can be used as a therapy accompanying diagnostic tool. It does not require the person skilled in medicine to use acupuncture teachings, because it is ideally suited for therapy control in many other therapeutic methods such as regulation therapies, bio resoncance procedures, homeopathy, light therapy, massages, physical therapy, cranio-sacral therapy (osteopathy) and treatments with alkaline minerals.

From the perspective of the ETD it should be the goal of whatever kind of therapy to transform the radiation quality of the degenerative toward the toxic and from there to the endocrine or normal radiation.

The ETD also provides a description of various individual phenomena that can occur in a Kirlian picture of the terminal points.

In some pictures of finger tips one recognizes that some discharge channels which are emerging in the Kirlian photography do not extend radially outward but are instead bent on curved paths towards the fingertips. Such phenomena are described by Mandel in his ETD as "cramp claws", other ring-shaped discharge channels around the fingertips he named "stress rings". Sometimes in Kirlian photography of a terminal point occur radiations on which the shadows of the

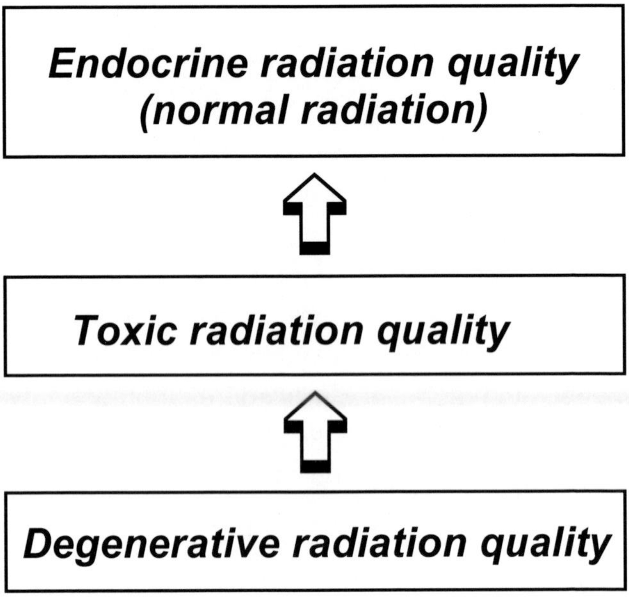

```
┌─────────────────────────────────┐
│  Endocrine radiation quality    │
│      (normal radiation)         │
└─────────────────────────────────┘
                ⇧
┌─────────────────────────────────┐
│    Toxic radiation quality      │
└─────────────────────────────────┘
                ⇧
┌─────────────────────────────────┐
│  Degenerative radiation quality │
└─────────────────────────────────┘
```

Figure 4: Optimal therapy path according to the Energetical Terminal Point Diagnosis (ETD)

finger- or toe-tips cannot be seen, but the picture is completely filled with radiations. Since this phenomenon occurs only in degenerative radiation quality, these types of emissions are referred to as degenerative plaques.

On many Kirlian pictures of the finger- and toe-tips gaps in the emanations are recognizable that are of course energetically interpreted as well and can be associated with a lack of energy.

On the basis of many thousands of patient images energy-organ-relationships show empirically strong correlations. Thus, the energy state of an organ is showing preferably in a particular sector of the emission of a finger or a toe. The resulting determined allocation of organs or body areas to certain meridians or terminal points is referred to in the ETD as an energy-organ-relationship. With this it is possible to locate and characterise the energetic situation of an organ in the Kirlian picture.

The darkroom effort is unsatisfactory for the user, because it has to be operated with the classic Kirlian devices to create an ETD of a patient. Only the development of digital image recording should change this, which will be further explained in the next chapter.

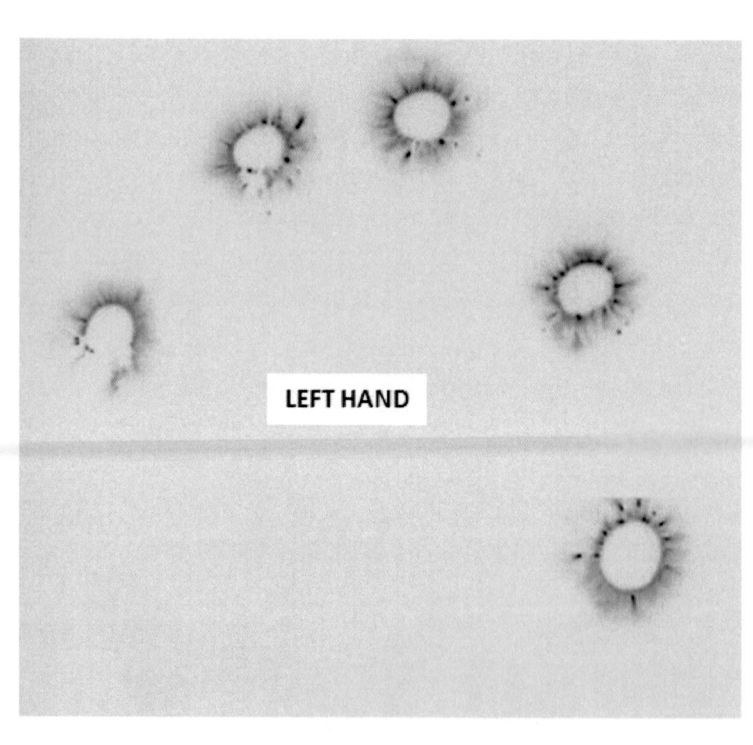

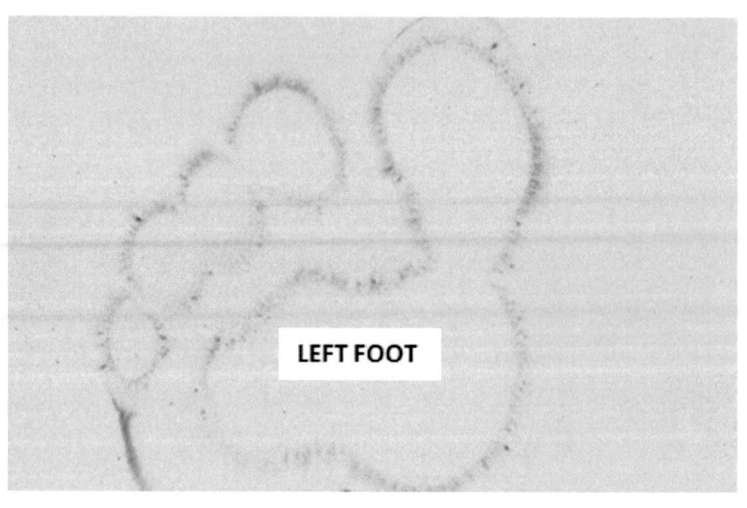

LEFT HAND

LEFT FOOT

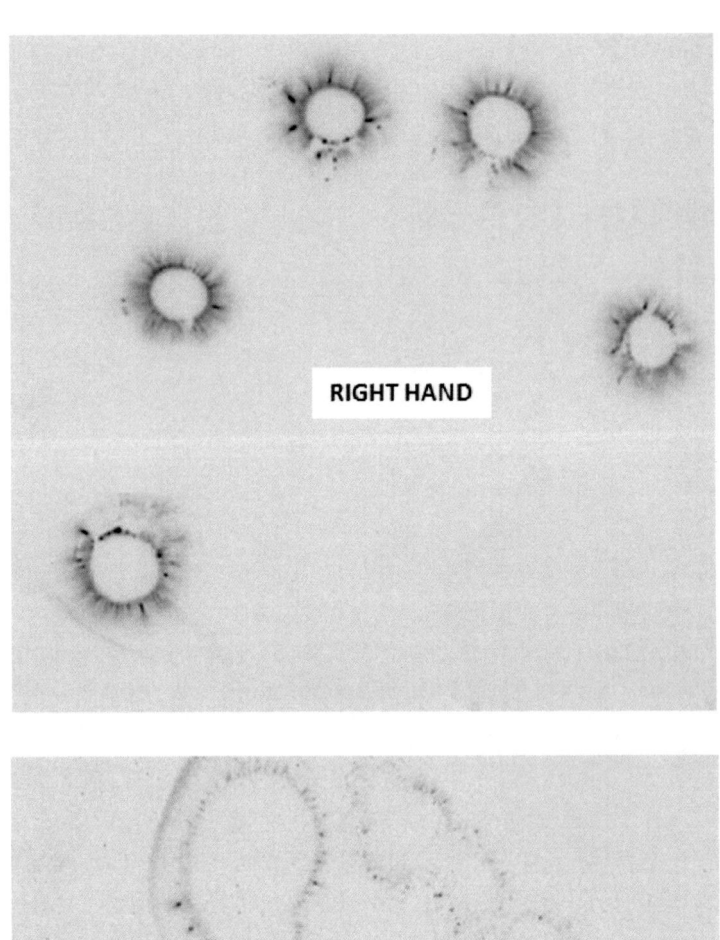

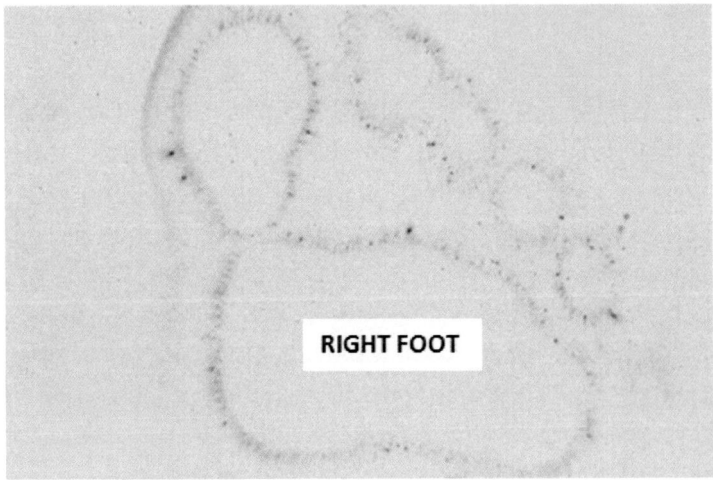

Figure 5: Typical Kirlian photograph of the terminal points

CHAPTER III

First personal experiences with the Kirlian Effect

In the 1970's, I stood at the beginning of my physics and mathematics studies in Münster, North Rhine-Westphalia. However, I was led to this study originally by my enthusiasm for astronomy, because the world of the stars reveals to man a mysterious inner harmony and order, which has fascinated me all along.

Like this classical mechanics celebrated, as a fundamental branch of physics with the formulation of the gravitational act by Isaac Newton their first major successes in the 17th century with the description of the track of motions of the stars as they were verifiable with former observation instruments at the time. Everything we have found out so far about the nature of stars, is ultimately derived from the analysis of the light emitted or reflected by them. The telescopes which people improved and enlarged more and more over the centuries thus serve primarily as light collectors.

Then again, physics advanced significantly

in the 19th century as their scholars decrypted the nature of light and recognized the visible light to be a small section of the spectrum of electromagnetic waves. Parallel to this discovery, physics made equal progress with the explanation and use of electric and magnetic phenomena and this is how the classical electrodynamics were formed as another important branch of physics and increased her reputation as queen of the natural sciences.

Especially in the first semesters of the studies in physics, these two sub-disciplines of physics – mechanics and electrodynamics – stand in the focus of the educational content and any reasonably dedicated physics student looks with great awe and wonder at the possibilities which are thereby provided for an exact description of nature.

As a matter of fact physics, especially classical physics deals only with objects of inanimate nature. This doesn't change initially, even if you consider the modern branches and extensions of physics, in particular the theory of relativity of Einstein and the atomic physics and the – for their formulation – indispensable quantum mechanics.

After studying modern physics more closely, it struck me as downright amateurish that in the established sciences it was mainly left to the disciplines of biology, medicine and chemistry to explore the nature and the laws of life. To explore the nature of stars, other than the range of visible light regions of the electromagnetic spectrum, had long been examined at the time of my studies.

In recent years the radio astronomy has developed quickly to become an important branch of astronomy, At the end of World War II british radar technicians discovered that even though objects in the sky, particularly the sun, radiate energy at many frequencies of the eletromagnetic waves and that you can, of course, watch and analyse these signals as well as the visible light, it seemed to me like a taboo in modern physics to refrain from the investigation of the living, even though she already had the deepest insight into the structure of matter and like in any other natural science, possessed the required model notions and the scientific equipment with which one could gain more knowledge about the jungle of the unknown life.

In later semesters, it was also common to choose a module as a student in the course of physics at my university. This choice suited my interests and would later be of benefit to me in my profession. So, my first choice fell again on astronomy, as it corresponded to my original love for her and with my interest in this science. But it soon became clear to me that astronomy within the department of physics at my university was not as attractive for me as I had hoped initially. But luckily there was still an alternative.

Besides studying, I had a job in a hospital, where I had previously done my community service. While working there, I was able to get myself a good insight into the various clinical departments and this offered me the possibility to bring the experiences gained there into my studies of physics and to look for synergies.

Therefore, I changed my optional subject and finally opted for the medical physics. This proved to be a very positive decision because the Insitute of Medical Physics was an insider's tip among the advanced physics students, as it had an ultra-modern instrumental equipment. In particular, in the fields of electron microscopy and atomic micro analysis of biological samples his former institute director, Professor Gerhard Pfefferkorn, was looked upon as an international luminary in his field, who had worked in his early years as an assistant to the famous Max von Ardenne in Berlin. Thus, from this man one could learn a lot.

At this time, it was not yet clear to me which career direction I would take later as a physicist, but to further explore and research the mystery of life was of great interest to me. I was especially interested in physical measurements with which the vitality and mental state of living beings could be characterized. I already suspected at the time that the eletrical and electro-magnetic properties of people stand in conjunction in a special way with their vital functions and their awareness.

Since I had no opportunities in the University environment to do research on these things in profoundness, I decided, in order to satisfy my curiosity on these issues, to set up a small laboratory at home, to make electrical measurements on plants at first. Of course, my financial resources were limited as a student and I didn't have money to purchase expensive measuring instruments. At that time, however, I already had a good basic knowledge of electronics and so I created my first electronic circuits, power supplies and measuring equipment myself. An old black and white TV was converted without hesitation into an oscilloscope, and since I had a good connection to the administrative head of the hospital I worked in, I could supply my lab with an old decommissioned ECG machine at a certain point, which was still fully functional and could do the same with other interesting equipment.

In this way, I started with the first electrical measurements on indoor plants and discovered to my amazement that they were very sensitive to people who approached them. On the signal wave shape that my converted ECG device recorded, I could even notice differences in the eletrical response of the plants to different people. For these measurements I also had to build a large Faraday's cage, made of wire mesh, to protect them from the ever-present interfering signals of the household eletrical network. For, every electrical socket, light switch and ceiling light and even the concealed power cables radiate the infamous 50-hertz hum, which is disturbingly noticeable in sensitive electrical measurements.

Sometime around this period, about 1980-1981, I became aware of the Kirlian effect. In a darkened room one could observe pale bluish emmissions of the objects recorded with a Kirlian device.

For inorganic, but electrically conductive objects these radiations were always constant and unchanging, but once you put your hand on the cover plate of such a device, you noticed changes in the emanations that were apparently dependent on the form of the day or the mood of the test person. I was literally electrified when I made initial experiments with this technique. These experiments were made possible to me by a friend who made his Kirlian device of dutch origin available to me for some time. Moreover, you could even then record these luminous phenomena on a black and white photo positive paper by placing it between the investigated object and the top plate of the Kirlian device. Then, the paper was exposed for a few seconds under darkroom conditions and subsequently developed in the customary manner, fixed and dried. From my school days, I still had the necessary photo laboratory equipment, to undertake extensive experiments with it. These models of Kirlian devices were already availale for purchase in these days.

The theme of "Kirlian photography" had at the time a certain popularity because interesting articles about it were appearing here and there in known magazines.

Finally, even an electronic circuit for the replica of a Kirlian device was published in the German electronic journal "Elektor" and this was exactly the challenge I had been waiting for, because it was now easy to build my own device. In order to shield the resulting high voltage outwardly,

I used a housing of commercially tailored acrylic sheets, through which the interior of my Kirlian device remained visible from the outside.

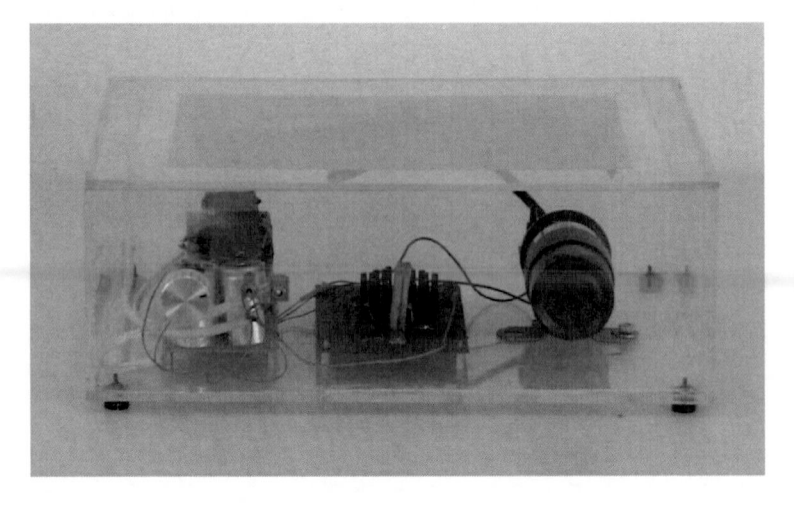

Figure 6: First selfmade construction of the 1980s

On the picture you can see the case of acrylic glass. Inside is the circuit board of the electronic circuit. From this, a high-frequency circuit provides a signal which is then transformed by an ignition coil which acts as a Tesla generator to the desired pulsed high voltage. A high voltage cable connects the coil with a sheet metal electrode of aluminum sheet which is attached under the top plate of the housing.

I had the unit in practical use until the end of the 1980s and it still exists today as a private museum piece. At first, my experiments were limited to the examination of plant leaves. I was particularly fascinated by the so-called phantom image effect. In literature there were circulating mysterious Kirlian pictures of plant leaves at the time. The photographs show the emissions of an unbroken leave, even though a part of this leaf had been cut off before. The esoteric explanation attempts did not impress me very much, but the phenomenon image attracted my scientific curiosity. After some attempts I also finally succeeded in reproducing such "phantom images", however, this effect turned out to be an artifact. Indeed, if laying the leaf in the intact state on the photo paper to be exposed, the aqueous outlines of the leaf will be precipitated under certain atmospheric moisture conditions on the photographic paper. If one then cuts off a portion of the leaf with a sharp utility knife, without changing the position of the object, the ominous phantom image is produced during exposure, because the deposited liquid on the photo positive paper is electrically conductive and leads to the exposure of the photo paper. So therefore, the phantom image effect is nothing more than a clever slight of hand.

After these first experiences, human test subjects moved into the foreground of my interest. In 1981 I became aware of this already; in the previous chapter I presented the Energetic Terminal Point Diagnosis by Peter Mandel. It was very impressive to me that one could find out empirically that the emissions of the finger- and/or toe-tips of a patient correlated with the organs inside the body.

This of course raised a lot more questions for me and in order to resolve these a deeper understanding of the electromagnetic conditions in the interior of the human body was necessary. In1982, however, the time of my theoretical diploma exam approached and due to the now upcoming exam preparation I had no more time left for dealing with these issues.

After my theoretical final examination I had to make another decision. In which institute and with which professor and on what topic should I write my practical graduate work? In this final phase of study for a physics degree, the student should prove through the work on a physical subject theme his ability to use the knowledge acquired during his studies. This is done by building experimental devices, carrying out a series of trials, evaluating the results, developing conceptual models from this and, explain them in mathematical formulas as far as possible.

My choice fell finally on to the Institute of Applied Physics and within this on to the department of the then private lecturer Dr. Kassing, who was later a professor at the University of Kassel and headed the Institute of Technical Physics there for many years. His department in Munster dealt with questions in the field of semiconductor physics. Under semiconductors, such as silicon, one understands in solid state physics substances that are taking on an intermediate state with respect to their electrical properties, between electrical conductors and insulators.

In microelectronics these semiconductors are of outstanding importance. The institute was particularly proficient in the use of microprocessors, because at the time the computer technology was in a revolutionary process of evolution and the first home computers had come onto the market. Through the mediation of Mr. Kassing I came to Siemens AG in Munich in early 1984 where I first worked on my diploma thesis and later on my doctorate rer. nat. There I was engaged in the development of microelectronic circuits, particularly in memory chips. Quantum effects play more than an insignificant role in this. I developed a design model with which one could simulate the electrical behaviour of components on a microchip. Through such simulations valuable time was saved in the development of new memory chips.

Moreover in the mid-1980s, when I was busy with my PhD, I was dealing privately with the Kirlian photography again. In my apartment in Munich I extended my recording options to also include color positive papers. During this time I also undertook an intense personal transformation process, which affected several areas of my life. I intensified my meditative side, which led to a strong development of my energy system. A conversion of my eating habits has also helped, because since that time I have not eaten meat and in the last decades I have tried variations of the vegetarian diet over periods usually spanning several years.

a)

August 1983

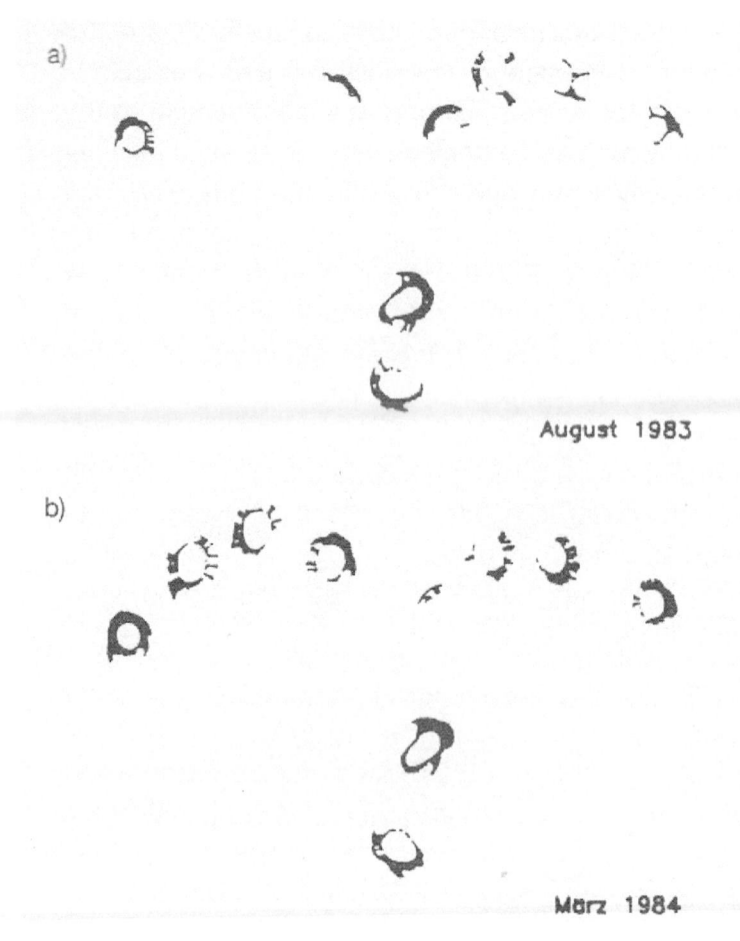

b)

März 1984

Figure 7: Documentation of a spiritual transformation process

From a lacto-vegetarian I then became a vegan – with a growing feeling of well being. I combined each transition to a new diet with various fasting techniques. Finally, I have also undergone a longer "boiling-

c)

August 1984

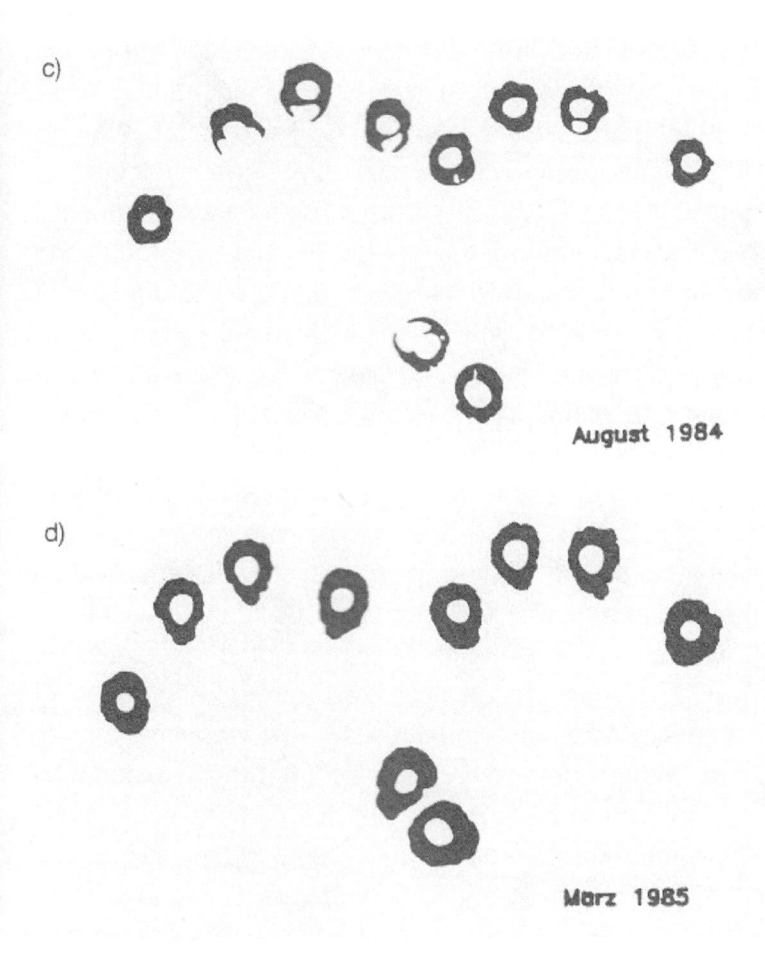

d)

März 1985

potless" phase with uncooked vegetarian food. This brought me a noticeable physical rejuvenation effect. For thirty years now, I have eaten a completely vegetarian diet and can find no deficiency symptoms or other restrictions.

I documented the most intensive development phase of my spiritual transformation with Kirlian photography. This took place between August 1983 and March 1985. In 1983 I was still in Munster and since the beginning of 1984 I lived in Munich. Through effective methods I managed to increase the energy that flows through the body enormously, which is reflected in the photographs of figure 7. Extraordinary spiritual experiences were happening in my life at that time. I got more and more access to memories deeply hidden in my soul. Reincarnation experiences accumulated during this period and its relevance became a personal certainty to me. My beliefs were the same with regard to the existence of the human chakra system. My individual experience of the energy increase resulted in a sustained manifestation of a state of uninterrupted happiness in myself. This was made possible by the strong sense of the never ending streams of energy and was measurable. Since then, it has been possible for me to experience arbitrarily in time prolonged periods of pure awareness, where my thought process becomes perfectly still. In this state, that is free of any thought process, I manage to be aware of a strong and constantly noticeable energy flow in my body. Since that time, I can measure consistently closed ray coronae of the terminal points of my finger- and toe-tips. This experience shaped the past thirty years of my life since the end of my studies and is also associated with a high level of health. Doctor's visits are therefore rare for me.

I have described the body-oriented and psycho-spiritual methods which led to a sustainable development of my energy system in the booklet The Quantu Temple, published by Scorpio, Munich a few years ago. For many years now, I teach these methods mainly to people from healing professions – among them are doctors, medical practitioners, as well as interested lay people. They complete the continuing education seminars organized by me in Quantum Practice ®.

I was especially interested from then on, on how to describe these processes that I had experienced with my own mind and body in physical models. Given present results of measurements it is becoming apparent that the intensity and quality of Kirlian photographs of the terminal points is connected with the vitality and the quality of consciousness of the person examined. Consequently, the health and state of consciousness of a human being must therefore be related to the electrical conductivity, ie the distribution of positive and negative electrical changes in the body.

Therefore, the question arose as to whether the individual radiation phenomena that can be found time and again in Kirlian photographs can be assigned to certain charge polarities, thus to electrically positive or negative charge carriers. This was a prerequisite to create physically based models of the Kirlian effect. Was there a plausible physical explanation for the normal radiation quality described by Peter Mandel in his Energetical Terminal Point Diagnosis (ETD) and also for the toxic and degenerative radiation qualities? For this it was necessary first of all to look at the exact time profile of a high-voltage signal in a Kirlian device. From this one can deduce the running direction of the carriers as they move through the air from the photographic subject to the high voltage electrode.

By using an oscilloscope the time laps of the high voltage on the electrode can be recorded well. Usually, Kirlian devices produce a so-called alternating field at the high voltage electrode. Although here, sharply defined voltage-peaks or -flanks are produced by the electronic circuit of the high voltage generators, these signals

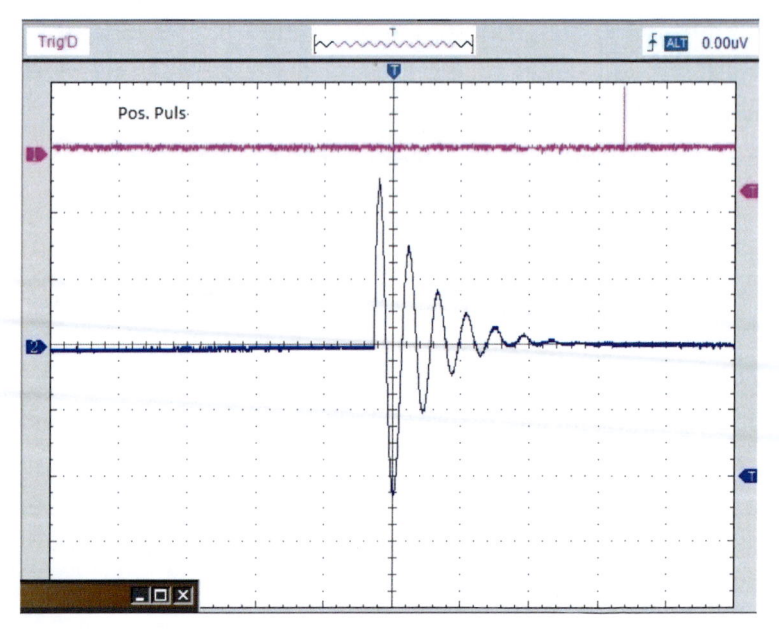

Figure 8: High voltage signal of a Kirlian device

are distorted by induction and mutual induction in the Tesla generator to a sinusoidal, exponential decay curve and only then get to the high-voltage electrode. Through this results an ever polarity alternating electric field at it, in which charged particles are accelerated first from the object to the electrode and then from the electrode towards the object.

In figure 8, the typical course of the high-voltage signal is illustrated. If an electrically positive high voltage pulse is set to the electrode, electrically negatively charged particles, that escape from the object to be photographed, will be accelerated towards it and produce streaks of light in the air through the collisional excitation processes already described. On the other hand, if there is an electrically negative high voltage pulse set to the electrode, then there are positively charged particles accelerated towards it which are emanating from the object recorded and generate respective light traces. Normally, positive and negative pulses are constantly changing at the electrode, as figure 8 shows. Therefore, in a normal Kirlian image the luminous traces of electrically positive and negative charges are superimposed. However, if one introduces a rectifier between the Tesla coil and the high voltage electrode, one can thus, depending on the direction of the rectifier, make it so that only the electrically positive or negative voltage pulses reach the electrode. I was able to realise this with high voltage resistant gas diodes. In the same way, Kirlian pictures can be generated so that they only record the luminous traces of negatively charged particles or the luminous traces of electrically positively charged particles. With the help of the rectification of the high voltage pulses, I was able to assign the radiation qualities known from the ETD unambiguosly to the electrical charge carriers of a certain polarity. The results are by no means a surprise, but do confirm the experiences of people skilled in medicine in relation to the role of electrical charge carriers in a biological organism.

So it turned out that the fine tuft-like and extra long discharge channels occuring during normal radiation are clearly assignable to the negatively charged electrons. Thus, normal radiation is produced only when the high-voltage electrode has a positive polarity, because only then will negatively charged particles move towards it. Since the normal radiation, especially if the coronae rays are harmoniously closed at the terminal points, only occurs in healthy patients, the presence of a sufficient concentration of mobilizable electrons which are free on in excited states, correlates obviously with vitality and health.

I could however clearly assign the phenomenologically described toxic radiation quality in the ETD to the point-like condensations of the discharge channels occuring in the Kirlian picture to positive charge carriers, because this radiation quality only occurs when a negative voltage is applied to the high voltage electrode. Since the toxic radiation quality according to results is correlated with inflammatory processes in the organism, it is mainly composed of positively charged molecules which occur frequently in the body tissue within an acidic medium. The positively charged ions of course include protons. In the so-called degenerative radiation quality of the ETD, the situation is aggravated, because here the subject recorded seems to be inundated by positive charge carriers, as is typically the case in overly acidic patients with severe metabolic disorders. Here also the dreaded free radicals reach maximum concentrations which will lead to a collapse of the organism in the long term.

The inner aureole at the terminal points contains both positively and negatively charged ions. Here, there are larger electrically charged molecules which cannot move too far from the object recorded in the alternating field of high voltage, due to their mass.

Through these studies with selectable polarity of the high voltage electrode, I had succeeded in taking an important step on the way to a deeper and more physically sound understanding of the luminous phenomena occurring during the Kirlian effect on biological objects.

But the crucial question remained open: Where does the light come from in our cells that had been tracked by Fritz Albert Popp and his colleagues? How has it entered here, how is it stored within and how does it contribute to the release of electrons and to the formation of further electrically negatively charged ions, which are required in the metabolism of a biological organism, since the optimal environment of the cell is basic?

CHAPTER IV

Without Biophotons there is no Bioplasma

Soviet biophysicists introduced the technical term bioplasma which found only little acceptance in the west. The secret Kirlian research in the USSR had long brought the scientists there to the conclusion that the concentration of free and easily mobilizable electrical charge carriers, in particular the negative ions, contributes significantly to the optimization of metabolic processes.

Expressed in the language of quantum physics: The electromagnetic energy stored in the body in the form of excited molecular and atomic states, with easily mobilized electrons, is the lubricant for the metabolism. This is the juice of life, a precious nectar, whose availability to the body of an organism depends from its eating, drinking and smoking habits, as well as on its physical stress and its emotional and mental state.

Why are especially the electrically negative charged ions so important for health? Because they have an anti-oxidative effect. They help the body to achieve a balance by neutralizing all the acid constituents of our metabolism or at least to ensure the maintenance of the basic character of the cells´ interior. As long as our body has a slight excess of negative ions, there is no threat of degenerative processes that push the aging of cells and thus the decay process of the body. In each cell membrane, which encapsulates the interior of a cell, small pores of specialized protein molecules are located to ensure that the cell remains slightly negatively charged relative to its external environment. This mechanism is known in biochemistry as the sodium-potassium pump and is shown schematically in figure 9. Whilst the pump is operational, three positively charged sodium ions are transported from the interior to the exterior of the cell, and during the subsequent reversal process, two positively charged potassium ions are transported from the exterior to the interior of the cell. The net result of the process is that the electrical charge of the cell is increased by one unit of negative charge for each complete operation of the pump. For the discovery of the potassium-sodium pump in the 1950s, the Danish physician Jens Christian Skou was awarded the Nobel Prize in Chemistry in 1997. The discovery of the sodium-potassium pump led the humanity to understand that its function is to protect the cell against excess acidity and helps it to maintain an optimum level of electrically negative charge against its environment.

Of course, this process, which is of fundamental importance for the survivial of the cell, has its limits. Because the concentration of positively charged ions in the surrounding area of the cell can exceed a certain value, the sodium-potassium pump cannot prevent the flow of positively charged ions back into the cell via diffusion inevitably

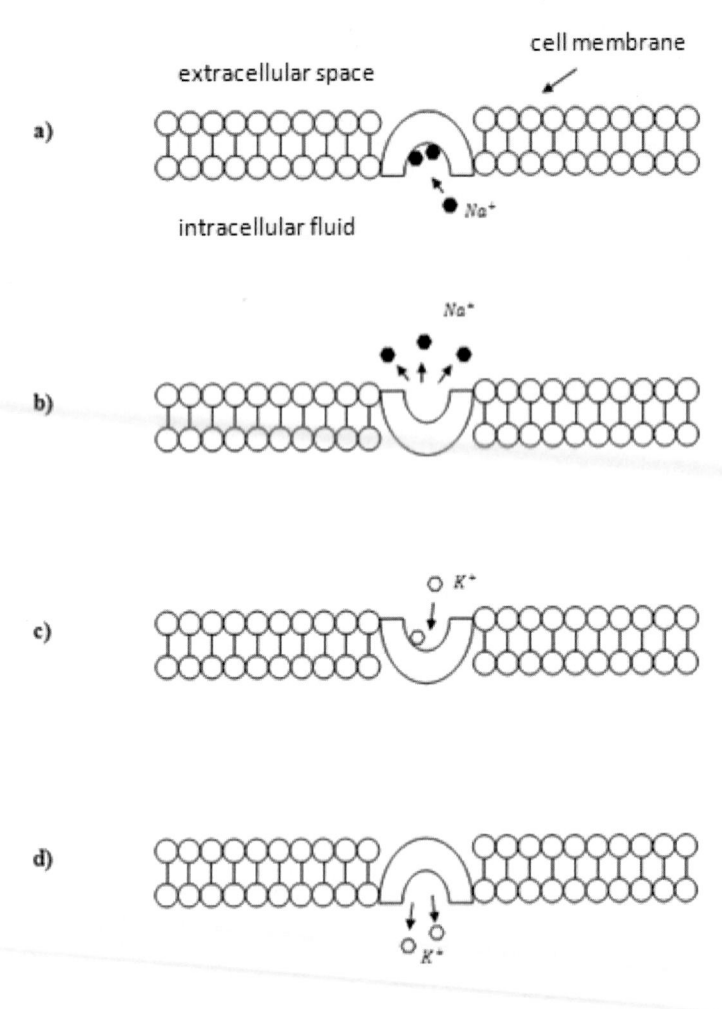

Figure 9: The phases a) to d) show the transportation of charged ions across the cell membrane using the sodium-potassium pump

Figure 10: State changes in atoms through absorption and emission of photons

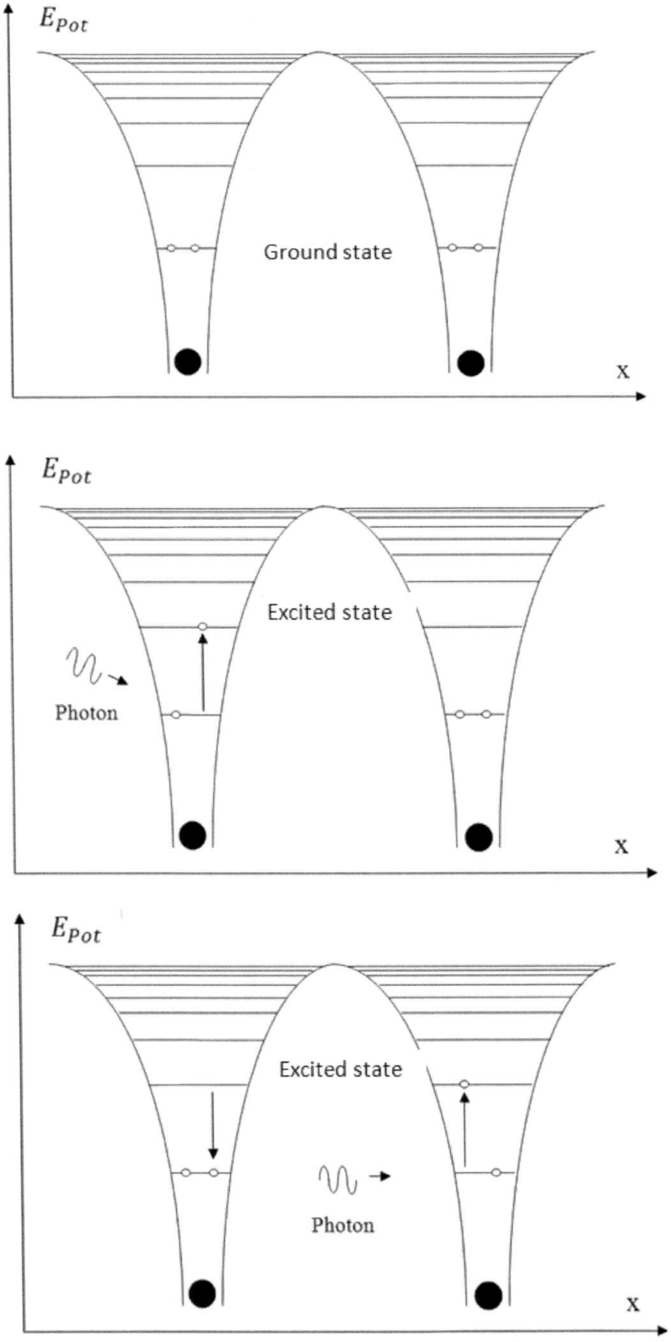
E_{Pot}
Ground state
x
E_{Pot}
Excited state
Photon
x
E_{Pot}
Excited state
Photon
x

leading to a breakdown of intracellular metabolic processes. Therefore, all patients whose Kirlian pictures have degenerative radiation quality are in a dangerous state for their organism because of acidosis which can leads to severe longterm disorders.

The energy to drive metabolism, which includes the operation of the sodium-potassium pump, is provided to the cell by ATP (adenosinetriphosphate) which is produced in its mitochondria and from basic raw-materials in the form of carbohydrates, proteins and fats supplied by the cell. ATP serves most biological organisms and is recognised as an universal energy source. In order for the complex metabolic processes in every healthy cell to work correctly and not end in chaos. Each cell requires an internally controlled electromagnetic field. This field is set up in the cell by the DNA in the nucleus and is distributed within the cell membrane. This is part of the discovery of the ultra-weak cell radiation.

The simple quantum mechanical model of Figure 10 illustrates the process of how the ultra-weak cell radiation is generated and made available for intracellular control processes. Shown here is the Coulomb potential of two atoms that is produced by the electric field in the vicinity of the positively charged nuclei of two atoms. The different energy levels inside the atom are recognizable and can be occupied by a single electron. In the upper part of Figure 10, both atoms are in the so-called ground state. In this case, all electrons reside on the respective lowest energy level within the atom. In the middle part of Figure 10 a photon coming from the outside passes through the atom. During its passage it can be absorbed by one of the electrons of the atom. For this absorption to occur the photon must have the minimum energy which is required for an electron to pass on to a higher energy level. After the absorption the atom concerned is in a so-called excited state. Most of the excited atomic states have only an average life span of a few millionths of a second. Thereafter, the electron jumps back from the higher energy level to the energy level of the ground state, whereby it emits a photon again to the outside to balance the energy difference between the two states. The lower part of Figure 10 shows the case where the radiated energy is again absorbed by an electron in an adjacent atom.

The process in the lower part of Figure 10 can be reversed in principle, as the ultra weak cell radiation is a coherent electromagnetic radiation. This means that the elctrons of different atoms in a biological cell can exchange photons back and forth with each other without the absorbed energy escaping anywhere – analogous to a ping-pong game in which both players try to keep the ball on the table for as long as possible. Hence, considerable amounts of energy in the form of excited atom- or molecule-states can be stored in a biological cell which is necessary for maintaining a controlled electromagnetic field within the cell. Using measurements, biophysicists have found that cells which were damaged by a mechanical impact, ie one in which the cell membrane was ruptured, lost significantly large amounts of light energy, due to increased emission unlike the case with undamaged cells. After the loss of the electromagnetic energy, stored in form of excited states in the cell interior, the cell dies as the cell metabolism that was previously controlled through this comes to a stand still. Measurements have shown that the majority of the electromagnetic energy stored in the cell is located in the nucleus. The nucleus is mostly made of the DNA (deoxyribonucleic acid) which is not only the carrier of the genetic information but also an electromagnetic store.

The DNA has the form of a plait-like multi-entangled giant molecule. If a sufficient number of electrons of the DNA are in excited molecular states, the electrical conductivity of the DNA increases. Consequently, many electrons can move freely back and forth within the DNA molecule. The electrons follow, due to the geometrical structure of the DNA, both quasi linear as well as circular tracks. Consequently, the DNA forms an ideal antenna due to its geometry. From the physical bases of the electromagnetism it is known that an eletrical charge that performs an accelerated motion, radiates an electromagnetic field. This allows the DNA to induce an electromagnetic field within the cell that serves to control the metabolic processes and the intra- and extracellular communication.

Each metabolic process represents a chemical reaction on the plane of atoms or molecules. From chemistry it is known that most reactions require activation energy to get started. This is partly achieved in the cell by the presence of the electromagnetic field which is generated by the DNA and is also available in the farthest corner of a cell. Nowadays, some of the very complex processes in the biochemical reactions in a cell are known. There are a number of proteins that break down or put together, like robots on a production line, other molecular components. These protein molecules then usually have a spirally twisted molecular component that can couple the energy from the electromagnetic field of the cell that they need for their activities.

Although these mechanisms in the cell interior have only been recently discovered, there was a similar-striking attempt at the macroscopic level in the early days of electrification. The famous inventor Nikola Tesla built a giant transmitter on the peninsula of Long Island near New York in order to radiate electrical energy wirelessly and to make it generally available. However, this was not in the interest of his investor of that time, J.P. Morgan, because he wanted to earn money with the sale of electric energy. After only a few years, this old idea was reattempted in order to supply small devices wirelessly with electrical energy. Inside of biological cells this has been working for hundreds of millions of years.

The electromagnetic energy stored inside of biological organisms is exchanged by photons and therefore these photons are referred to as biophotons. Pure physics considers however, that biophotons are no different from ordinary photons. The name biophoton only states that it is a photon inside a biological organism. Hence, the names bioplasma and biophotons are closely related because without biophotons there would be no bioplasma and vice versa. So electrical charges play a central role in the processes inside our cells and without the control of this charged balanced biological life would not be possible.

Biological life is a constant balancing act between order and chaos. In physics, this leads to the concepts of entropy and negentropy. In thermodynamics, a generalized form of the theory of heat, the entropy is a measure of the chaos – the degree of disorder. Experience has shown that in the inanimate nature all processes run towards chaos. For example, a hot cup of coffee will cool down more and more until the heat energy contained in the cup has evenly distributed itself on the surrounding space. This can be noticed if you have forgotten to drink a hot coffee and find after a while that only lukewarm or cold coffee is left. Usually the concentration gradient of positive and negative charges inside and outside of a biological cell would balance very quickly using diffusion processes, but here the sodium-potassium pump makes a charge separation which is a typical negentropic flow that counteracts the charge chaos.

In biological organisms there are internal fields allowing for the exchange of energy and information between atoms and molecules on the inside of the cell. In order to go to the bottom of the root cause of these fields, a deeper understanding of the construction and the structure of elementary particles such as the electron is required. Only transdimensional models such as the ones by the pioneers Charon and Heim or one of the various versions of the string theory can give further explanation here. The interested reader will find a detailed and easily understandable explanation of such transdimensional models in my book The Ancient Word – The Physics of God, which was published by Scorpio, Munich, 2010.

CHAPTER V

The Kirlian Effect on the Way into the Digital Age

As I knew from my time at Siemens AG, the development of powerful chips for digital recordings of photographs was still in its infancy at the end of the 1980s, but I was already sure at the time that procedures such as the Kirlian photography would develop further in a digital direction at some point.

But until one could achieve a resolution as good as that of the classic photographic method with the digital recording method, it would still take a while. First, completely different technical problems had to be solved in order to make the transition into the digital world. The place of the positive photo paper, which usually recorded the emissions of the object (for example, a finger or toe) had to be taken by a digital camera, and of course the objective requires a clear view of the receptive field. At that time there was a small workshop in Munich, which produced these so-called „Miracle Pictures". These consisted of two plane-parallel glass plates that were sealed at the edges with silicone and encased in a wooden frame. Between the sheets of glass, which had a distance of a few millimeters, an aqueous liquid was imprisoned with multi coloured layers of sand. If one turned the picture, the sand began to move slowly and produced magically moving images that were constantly changing and beautiful to look at. This gave me the idea for a transparent, high-voltage electrode. I then used two plane-parallel acrylic glass plates and filled the interior with an electrically conductive liquid. At one point I lead a wire between the plates and combined this with a high voltage generator and now my first transparent high voltage electrode was finished. Now one could position an object, for example a finger tip, on the one side of the acrylic plate and a camera on the other side with which one could photograph through the transparent high voltage electrode.

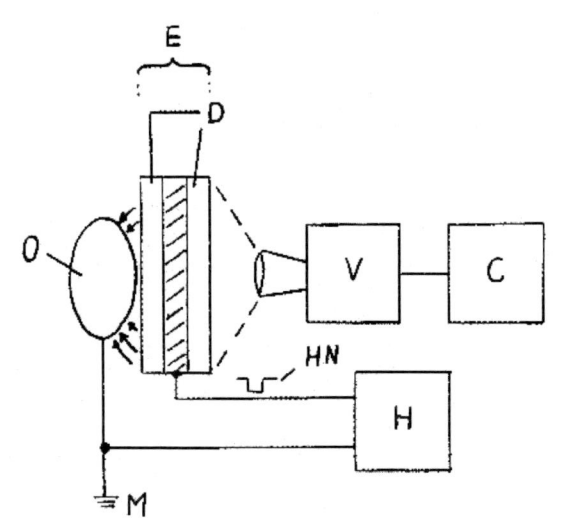

Fig. 11: Cross section through a Kirlian device with digital image recording

This even worked already with classic cameras. You only had to expose it for just a few seconds under high voltage to produce a Kirlian photograph.

Later, as soon as technically available, I replaced a traditional camera with a computer-controlled digital camera. With these ideas I registered a corresponding method at the German Patent Office and was then granted the patent in 1987. Figure 11 shows the schematic structure in an original drawing from my patent.

The object O is located on the transparent outer cover plate D which forms the high voltage electrode E with the inner transparent cover plate and the electrically conductive fluid located in between. Behind the inner cover plate the digital camera V is located that records the Kirlian picture electronically and forwards the electronic image to the evaluation computer C. In principle, the Kirlian devices of today with their electronic image recording still work in the same way.

Aside from the experimental test setup, I could not realize the digital part of the process technically at that time, because the digital techique had not yet been developed. Nevertheless, I had already achieved technical progress, by working together with a friend who is an electrical engineer. So a more stable housing was designed and the high-frequency high voltage generator was improved. With this generation of devices it was possible for the first time to vary the polarity of the high voltage pulses and to control the frequency and intensity and to set the exposure time automatically.

This especially facilitated the reproducibility of the images, because it was easier for the recording parameters to be kept constant. Of these devices at least one to two dozen came into circulation and were mainly used in medical practices.

Of course the recording process in itself remained unsatisfactory because you still had to constantly handle toxic chemicals in the development of positive photo papers and the development process required darkroom techniques at any location and was time consuming. I already described these circumstances in the description of the Energetic Terminal Point Diagnosis (ETD) earlier. So, I was still highly motivated to develop a market-ready digital Kirlian photography device.

After various applications for funding of the research and development projects of such a system, I finally came to agree a partnership with the Hans-Sauer-Foundation in Deisenhofen, near Munich. The founder, Hans Sauer, had made his name in the relay technology industry. He holds several hundred patents. He can also be found in the inventor gallery of the German Patent Office. The pioneer of space technology, Dr. Ludwig Bölkow, who was involved in the Board of Trustees at the time, supported the project as well. With the financial assistance of the foundation it became possible to develop several prototypes of the first computer-controlled Kirlian measuring station with electronic image recording in 1994. Some of these devices were also tested in practices of physicians and naturopaths and proved the usefulness of this technique for the complementary medical diagnosis. Due to the relatively low resolution of the digital camera used at the time, only photographs of individual finger and toe-tips could be taken in sequence. The first digital camera I used for these devices, a SBIG ST4, was originally developed for astronomical purposes. Therefore, it was also suitable for longer exposure times.

At this time the small transfer speeds between recording system and evaluation computer presented another technical obstacle for the practical application of the digital Kirlian technology. But the first system, which had been developed by me was equipped with an automated analysis of the recordings in connection with the control computer. For the first time, one was no longer solely dependent on human interpretation of the images, as the computer software did the qualitative and quantitative evaluation. In addition, the computer program was able to present the recordings of individual finger- and toe-tips together with an evaluation template on the screen and to then save and print it, which facilitated the recognition of energy-organ relationships and the comparison with other recordings.

I then extended my research to other biological and inorganic samples. Interesting results could then be gained from food and drinking water. The freshness of food correlates with the concentration of freely available or easily mobilizable electrons, as was to be expected. The spectacular result of the investigation of a water sample which was held by a person in her hands for one minute immediately before the measurement, makes it clear, that there can be a sustained exposure to electromagnetic energy through laying the hands on it. The first computer-controlled measuring device enabled programmatic controlled pictures of a series of measurements. In the process a previously manually treated water sample was exposed exactly every 60 seconds for every 3 seconds to the high voltage of the Kirlian device and thereby the CCD of the electronic camera was exposed. All individual images were then evaluated one by one by the computer to determine the intensity of the radiation corona on the sample container. The recordings were made at a constant sample temperature. Further external influence

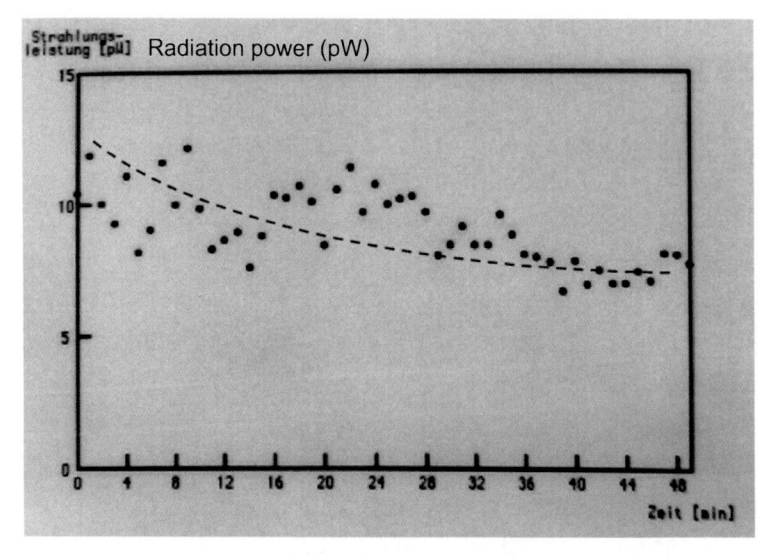

Fig. 12: Decay characteristics of a water sample after exposure to laying hands on it

factors can be ruled out as well, the picture being taken in a hermetically darkened room and also the recording parameters, such as exposure duration, frequency, pulse duration, intensity, time separation of the individual measurements were kept constant. Figure 12 therefore shows the time decay of the energy that had been transferred to the water by laying hands on it.

This computer controlled measuring device offered for the first time the possibility to quantitatively evaluate the obtained image data with a program, in particular the corona rays at the fingertips. I developed an analytical method in the mid-1990s, with which the corona rays can be systematically evaluated both qualitatively and quantitatively. Qualitative analysis means the classification of the radiation quality, wherein the phenomenological description known from the (ETD) proved to be useful. Normal radiation, toxic and degenerative radiation quality are clearly linked to the electrical charge ratios in the body and reflect the vitality status at a cellular level. The radial analysis, which I called the computerized scoring system, is equipped with routines that can distinguish from each other the three radiation qualities known from the ETD. First, in a radial analysis the center of the radiation corona is calculated. From this centre the intensity of the radiation corona is determined in each direction. By calculating the circumference of the radiation corona, the normal radiation can be distinguished from the degenerative radiation quality, because the wisp like topography of the radiation corona at normal radiation is to a much greater extent than in the degenerative radiation quality. The toxic radiation is well distinguished from the other two qualities, since its dot-like densities are much brighter than any other corona discharges. With this an important step towards standardized and objective evaluation was done because so far only a person trained in the ETD could make these distinctions. With the radial analysis it is also possible to assign the position of conspicuous anomalies to a certain angle and thus to a particular organ. Another advance is the ability of the quantitative evaluation because now also the total intensity of a radiation corona and the sum of all radiation corona can be determined.

The next generation of devices then brought a breakthrough both in the quality of the recordings as in their transmission speed to the computer. This system, which came on the market in early 2000, was developed under license of a medical technology company in the Lower Rhine area working together with me and this contributed to the further spread of the digital procedure. After all, with this system the fingers of one hand or the toes of one foot could be recorded all together at the same time in a satisfactory resolution.

However, the aim was still to develop a digital recording system that could accommodate all ten fingertips of both hands simultaneously with all toes of both feet at the same time as well. This made it necessary for an increase in the recording field and hence improved digital cameras. The small market niche for Kirlian devices had of course no effect on the further technical development and so another ten years had to pass until this goal could be achieved.

The current generation of new photon diagnostic equipment meets all the desired properties to make the transition into the digital age possible for the complementary medical diagnostics. With the new developments in this field, other technical problems have been solved too. So darkroom techniques are now no longer required for recordings of the patient because the devices also have an appliance for darkening the exposure field while recording and therefore, the recordings can now be made under normal daylight conditions.

CHAPTER VI

The Photon-Diagnosis

I call "Photon-Diagnosis" a measuring method with which the radiation of a biological object is photographically recorded and evaluated by a computer using the Kirlian effect. The method can be adapted to different living organisms, especially people. It is based on a recognized microbiological, biochemical and physics methods.

An important aspect in this is the execution of the measurements, which must be taken under reproducible conditions. The design of the measuring arrangement always provides constant recording parameters such as exposure time, high-voltage pulse, temperature, humidity and object-orientation. This allows for the comparison and contrast between different recordings under objective conditions. Therefore, the photon-diagnosis is particularly suitable for multiple measurements for therapy monitoring in the naturalistic and medical practice.

The photon diagnosis is primarily an imaging technique that provides information about the quality and distribution of the electric charge in the human body.

Fig. 13: Digital image recording of fingertips in Photon-Diagnosis

These allow a characterization of the general state of the health of this person. The evaluation program of the photon-diagnosis automatically provides a representation of the energy-organ-relationships which are based empirically on the accumulated experience of the ETD (Energetic Terminal Diagnosis). In addition, the images are evaluated quantitatively by the interpreting program of the photon-diagnosis, allowing for a comparison with average measurements.

Digital image processing has reached many applicable areas in recent years and partially revolutionized them. Technically, this is primarily due to the fact that relatively inexpensive microchips are now readily available, with which photographs can be taken in a quality, as it has been possible only with photographic emulsions, for example in a smaller or bigger film picture format than in the past. Almost every mobile phone is now equipped with a small digital camera that allows the user to take snapshots at any time without having to rely on the developing of the pictures taken in a photo shop. In addition, the images can also be forwarded with a few clicks on the control panel of the mobile phone over the internet. Even in areas of the professional photography, especially in reporting, digital photography has long begun its triumphant march. Some manufacturers of classic cameras had to learn about this the hard way, as their sales figures plummeted. If they failed to serve the rapidly growing demand for digital photography, this would have continued in a downward spiral.

In digital photography the image information is not buffered on the negative film but on a CCD (Charge Coupled Device). Such a CCD is placed on a microchip and is composed of several millions of individual photo transistors. Each of them forms a pixel, ie, an image point of the picture to be recorded. These transistors are regularly arranged in rows and columns. They collect the light falling on them during the exposure and transform it into an electrical signal which is converted into a digital image by an electronic circuit, which is then stored in the storage area of a microcomputer in a common image file format.

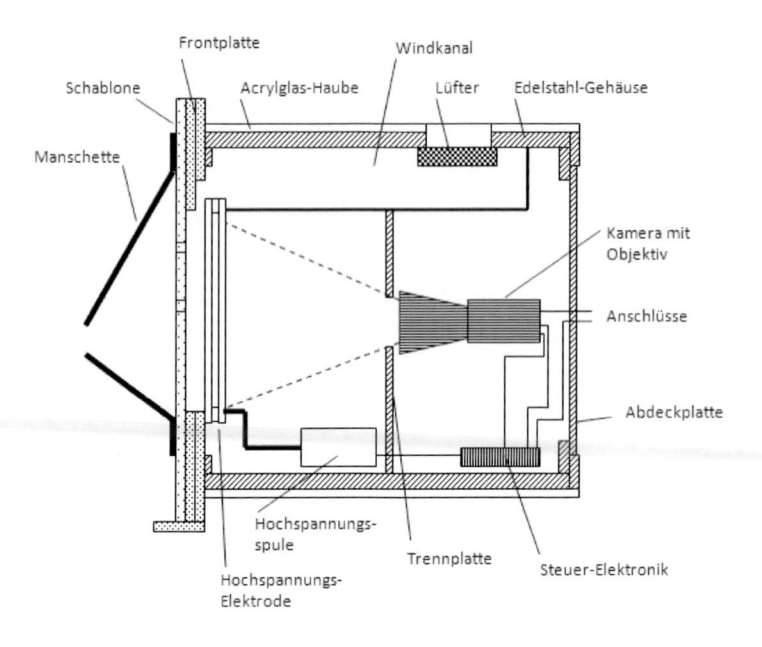

Fig. 14: Cross section through a modern Kirlian device for digital image recording in Photon-Diagnosis

If you want to do a simultaneous digital Kirlian recording of all fingers or all toes of a person, you need of course a very large receptive field and a digital camera with a good optical lens and a high resolution.

Although, this alone is not enough. As Figure 14 shows, modern Kirlian devices that record all terminal points, are also equipped with a cuff which makes it possible to take a picture in daylight conditions without the recording room having to be darkened. It is sufficient to use the sleeve in order to keep the daylight from the transparent high-voltage electrode. The person to be recorded put their arms or feet through the cuff and sets their fingers or toes on the front of the electrode and the recording can begin.

Before recording however, the digital camera is in a preview mode, that is, it constantly delivers images from the receiving field which is illuminated with LED's in order to be able to control whether the person has placed all terminal points on the plate. Otherwise, there would be the risk of creating artifacts during the actual recording lap of luminous phenomena, when the high voltage was switched on. In addition, modern devices are also equipped with a ventilation appliance in the recording field, as otherwise moisture could collect on the electrode from fingers or toes. This would inevitable lead to phantom images which could distort the measurements and complicate interpretation. For example, if a patient does not put all the toes on the electrode during the recording and this were to remain unnoticed, an artifact could result. Because instead of a rays corona there would appear at best a fuzzy patch or a toe could be entirely missing.

The finger tips can only be pushed through the openings of a mask on to the recording electrode. Through this, they are arranged in the same position and the same angle to each other for every shot. Thus, it is possible to represent the rays of the corona of each terminal point on the computer monitor with the respective energy-organ-relationship simultaneously.

Figure 15 shows an evaluated recording with the photon-diagnosis. Shown are the ten rays of the coronae of the fingertips and their assignment to

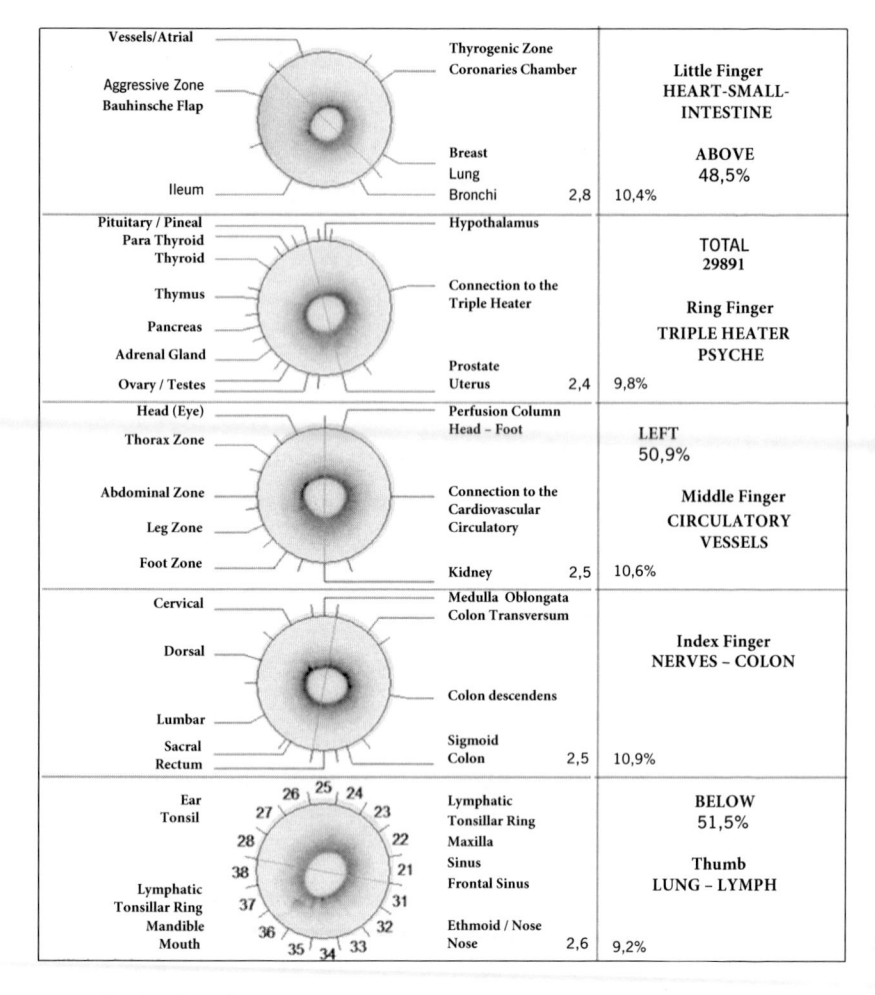

Vessels/Atrial Aggressive Zone Bauhinsche Flap Ileum	Thyrogenic Zone Coronaries Chamber Breast Lung Bronchi 2,8	10,4%	**Little Finger** HEART-SMALL- INTESTINE ABOVE 48,5%
Pituitary / Pineal Para Thyroid Thyroid Thymus Pancreas Adrenal Gland Ovary / Testes	Hypothalamus Connection to the Triple Heater Prostate Uterus 2,4	9,8%	TOTAL 29891 **Ring Finger** TRIPLE HEATER PSYCHE
Head (Eye) Thorax Zone Abdominal Zone Leg Zone Foot Zone	Perfusion Column Head – Foot Connection to the Cardiovascular Circulatory Kidney 2,5	10,6%	LEFT 50,9% **Middle Finger** CIRCULATORY VESSELS
Cervical Dorsal Lumbar Sacral Rectum	Medulla Oblongata Colon Transversum Colon descendens Sigmoid Colon 2,5	10,9%	**Index Finger** NERVES – COLON
Ear Tonsil Lymphatic Tonsillar Ring Mandible Mouth	Lymphatic Tonsillar Ring Maxilla Sinus Frontal Sinus Ethmoid / Nose Nose 2,6	9,2%	BELOW 51,5% **Thumb** LUNG – LYMPH

Left Hand

Fig. 15a und 15b: Representation of a recording of the fingertips in the photon-diagnosis

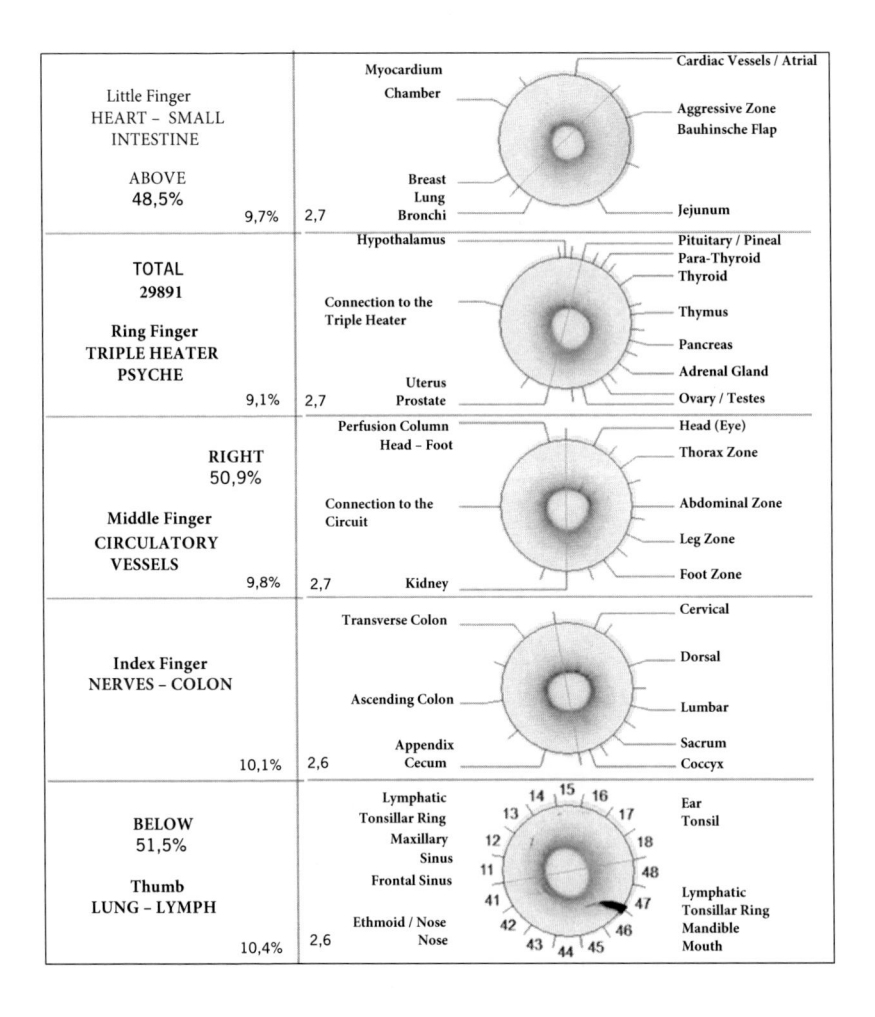

Little Finger HEART – SMALL INTESTINE ABOVE 48,5% 9,7%	2,7	Myocardium Chamber Breast Lung Bronchi	Cardiac Vessels / Atrial Aggressive Zone Bauhinsche Flap Jejunum
TOTAL 29891 Ring Finger TRIPLE HEATER PSYCHE 9,1%	2,7	Hypothalamus Connection to the Triple Heater Uterus Prostate	Pituitary / Pineal Para-Thyroid Thyroid Thymus Pancreas Adrenal Gland Ovary / Testes
RIGHT 50,9% Middle Finger CIRCULATORY VESSELS 9,8%	2,7	Perfusion Column Head – Foot Connection to the Circuit Kidney	Head (Eye) Thorax Zone Abdominal Zone Leg Zone Foot Zone
Index Finger NERVES – COLON 10,1%	2,6	Transverse Colon Ascending Colon Appendix Cecum	Cervical Dorsal Lumbar Sacrum Coccyx
BELOW 51,5% Thumb LUNG – LYMPH 10,4%	2,6	Lymphatic Tonsillar Ring Maxillary Sinus Frontal Sinus Ethmoid / Nose Nose	Ear Tonsil Lymphatic Tonsillar Ring Mandible Mouth

Right Hand

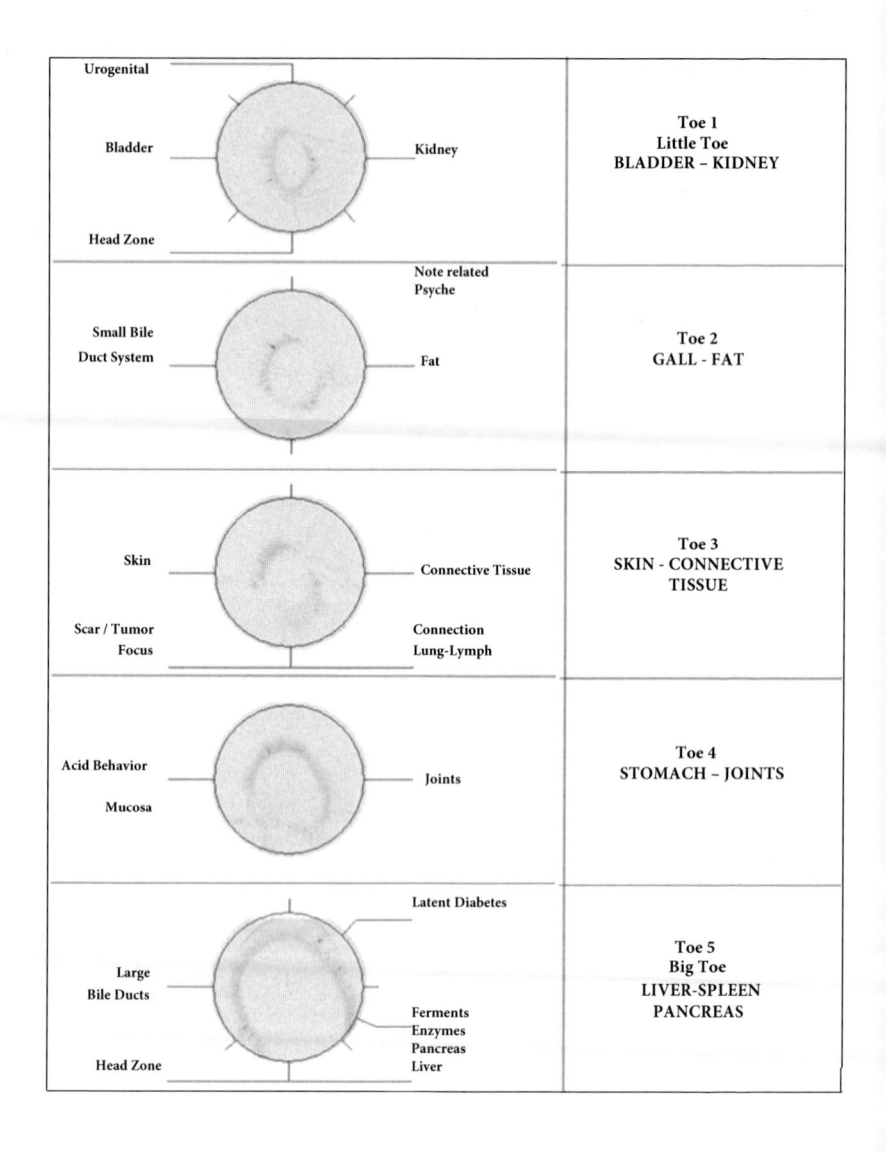

Urogenital — Bladder — Head Zone — / — Kidney	**Toe 1** Little Toe **BLADDER – KIDNEY**
Note related Psyche — Small Bile Duct System — / — Fat	**Toe 2** **GALL - FAT**
Skin — Scar / Tumor Focus — / — Connective Tissue / Connection Lung-Lymph	**Toe 3** **SKIN - CONNECTIVE TISSUE**
Acid Behavior — Mucosa — / — Joints	**Toe 4** **STOMACH – JOINTS**
Latent Diabetes / Large Bile Ducts — Head Zone — / — Ferments Enzymes Pancreas Liver	**Toe 5** Big Toe **LIVER-SPLEEN PANCREAS**

the Meridian System known from the TCM (Traditional Chinese Medicine). The image shows a nearly perfect radiation as it corresponds to the normal radiation quality in the ETD. According to the experiences with the ETD

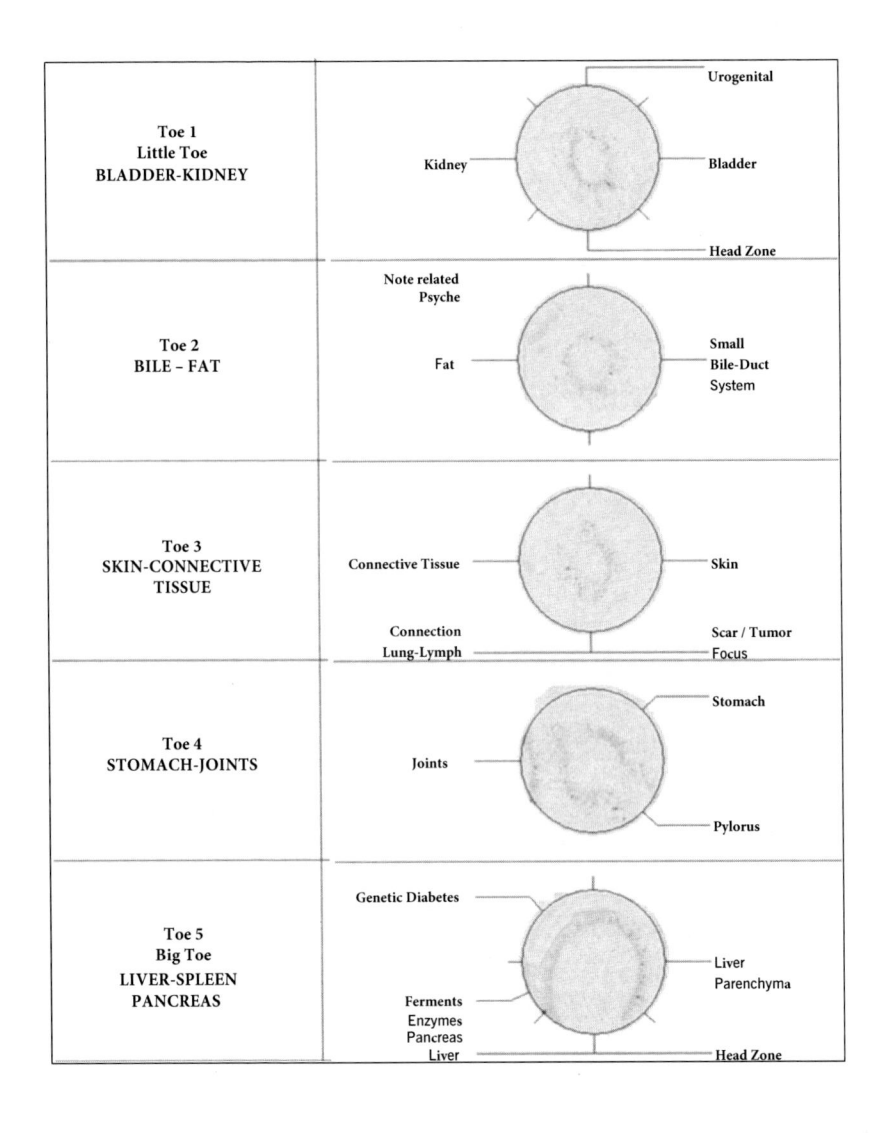

Toe 1 Little Toe BLADDER-KIDNEY	Kidney	Urogenital Bladder Head Zone
Toe 2 BILE – FAT	Note related Psyche Fat	Small Bile-Duct System
Toe 3 SKIN-CONNECTIVE TISSUE	Connective Tissue Connection Lung-Lymph	Skin Scar / Tumor Focus
Toe 4 STOMACH-JOINTS	Joints	Stomach Pylorus
Toe 5 Big Toe LIVER-SPLEEN PANCREAS	Genetic Diabetes Ferments Enzymes Pancreas Liver	Liver Parenchyma Head Zone

Fig. 16a und 16b: Display of an image of tiptoes in Photon-Diagnosis

the radiated emissions correlate in the sectors of the individual fingers with the organs that are in conjunction with the correspoinding meridian. On the two pages above are presented the five fingers of each hand in the following order: little finger, ring finger, middle finger, index finger and thumb. At the tips of the little fingers ends the heart-small intestine meridian. In case of energetic conspicuousness, the ray coronae of these fingers are showing the duodenum, the small intestine with jejunum and ileum and Bauhinsche flap, which makes the transition of the small intestine to the large intestine, as well as areas of the heart with artrial chambers, the hearts´ coronary arteries and the heart muscle (myocardium).

On the ring fingers ends the triple heater-psyche meridian. Here, in the various sectors of the fingers the energy states in the hypothalamus and in the endocrine glands are projected. These are the pineal, pituitary, thyroid, thymus, pancreas, the adrenal glands and the gonads which are the ovaries in a female and the testes in a male person.

At the tips of the middle finger ends the circulation-sex meridian. From the radiations at these terminal points there are indications on the energetical situation of the blood vessels system, whereby a distinction between the head zone, the thoracic area, the abdominal zone, the leg and the foot zone is possible.

The tips of the index fingers mark the end points of the nerve-colon meridian. There exists a relationship to the central nervous system in some sectors of these rays coronae with the subdivision in cervical (cervical) vertebrae, thoracic vertebrae (dorsal), lumbar vertebrae (lumbar) and the coccyx (sacral) and in other sectors an assignment to regions of the colon (ascending colon, transverse colon, descending colon, sigmoid colon).

Finally, the two thumbs contain information about the entire head area and the ear-nose- and throat region. The thumb tips are the terminal points of the lung-lymph flow. The sectors in the ray cornae of the thumbs give clues to the teeth, uper and lower jaw, mouth, nose, ethmoid, frontal sinuses and sinuses as well as the tonsils.

Figure 16 shows an evaluated picture of the toe tips of a person. On the left side in the radiation of the left toe, from the big toe above to the small toe down and on the right side accordingly the toes of the right foot shown with the energy-organ relationships as they are known from the ETD.

Toe tips also form the endpoints of certain meridians according to the acupuncture teachings, which lead to further empirically validated energy-organ relationships. So the tips of the big toe form the terminal points of the bladder-kidney meridian and from the radiations the energetic state of the genitourinary system with the bladder and the kidneys can be detected.

Furthermore, in the photon-diagnosis a quantitative analysis of the radiations is carried out. This gives the percentage by which each meridian is involved in the total radiation as well as the percentage of up-down and left-right distribution of the radiations.

In particular, during the recording of the fingertips, the fingers must be placed on the recording electrode slightly spread out, so that their coronae rays are not overlapping. Therefore, a definition of the axis angles is required. In Figure 17 these axes are shown in the diagram. They result from the position of the fingertips on the recording electrode. In assigning the energy-organ relationships in the photon-diagnosis these angles are taken into account. In Figure 17, the reference system for defining these angles in relation to the symmetry axis of each finger is shown.

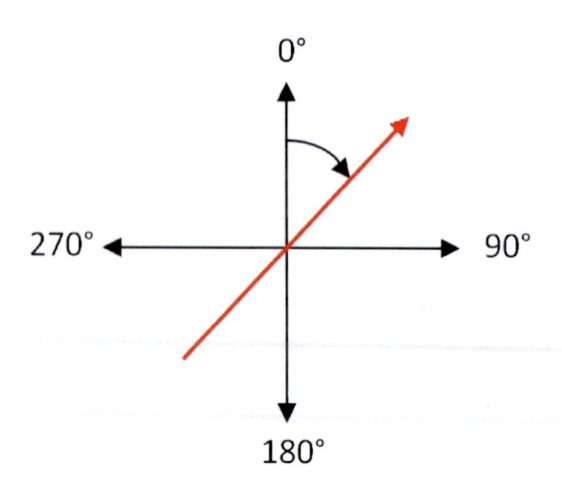

Fig. 17: Definition of symmetry axis angle

Energy-Organ-Relationships of the Fingers

Left Little Finger (Heart-Small-Intestine Meridian)
Axis Angle:	315°
Vessels, Atrial:	[0°–80°]
Thyrogenic Zone, Coronaries, Chamber:	[80°–180°]
Congestive Zone Lymph, Breast, Lung, Bronchi:	[180°]
Ileum:	[180°–295°]
Bauhinsche Flap:	[295°–0°]

Right Little Finger (Heart-Small-Intestine Meridian)
Axis Angle:	45°
Small Intestine, Aggressive Zone:	[0°–65°]
Jejunum:	[65°–180°]
Congestive Zone Lymph, Breast, Lung, Bronchi:	[180°]
Myocardium, Chamber:	[180°–280°]
Vessels, Atrial:	[280°–0°]

Left Ring Finger (Triple Heater – Psyche Meridian)
Axis Angle:	345°
Hypothalamus:	[10°–20°]
Horizontal Connection to the Triple Heater:	[20°–160°]
Uterus, Prostate:	[160°–200°]
Ovary, Testis:	[200°–220°]
Adrenal Gland:	[220°–235°]
Pancreas:	[235°–260°]
Thyroid:	[260°–280°]
Thyreoidea:	[280°–305°]

Para-Thyroid: [305°–335°]
Pituitary: [335°–350°]
PinealGland: [350°–10°]

Right Ring Finger (Triple Heater – Psyche Meridian)
Axis Angle: 15°
Pineal Gland: [350°–10°]
Pituitary: [10°–25°]
Para-Thyroid: [25°–55°]
Thyroid: [55°–80°]
Thymus: [80°–100°]
Pancreas: [100°–125°]
Adrenal: [125°–140°]
Ovary, Testis: [140°–160°]
Uterus, Prostate: [160°–200°]
Horizontal Connection to the Triple Heater:[200°–340°]
Hypothalamus: [340°–350°]

Left Middle Finger (Circulation-Sex Meridian)
Axis angle: 0°
Lymbic System: [350°–10°]
Note Horizontal and Diagonal
Connections: [0°–170°]
Kidney: [170°–190°]
Leg Zone, Foot Zone [190°–260°]
Thorax Zone, Abdominal Zone, Lymph 2 and 3: [260°–330°]
HeadZone: [330°–0°]
Eye: [340°–350°]

Right Middle Finger (Circulation-Sex Meridian)

Axis Angle:	0°
Lymbic System:	[350°–10°]
HeadZone:	[0°–30°]
Eye:	[10°–20°]
Thorax Zone, Abdominal Zone, Lymph 2 and 3:	[30°–100°]
Leg Zone, Foot Zone:	[100°–170°]
Kidney:	[170°–190°]
Note Horizontal and Diagonal Connections:	[190°–0°]

Left Index Finger
(Spine - Spinal Nerves - Large Intestine Meridian)

Axis Angle:	10°
Transverse Colon:	[0°–45°]
Descending Colon:	[45°–135°]
Sigmoid Colon, Aggressive Zone:	[135°–160°]
Rectum, Aggessive Zone:	[160°–180°]
Sacral:	[180°–205°]
Lumbal:	[205°–260°]
Dorsal:	[260°–300°]
Cervical:	[300°–340°]
Medulla Oblongata:	[340°–0°]

Right Index Finger
(Spine - Spinal Nerves - Large Intestine Meridian)

Axis Angle:	350°
Transverse Colon:	[300°–0°]
Ascending Colon:	[240°–300°]
Appendix Cecum, Aggressive Zone:	[180°–240°]
Coccyx:	[160°–180°]

Sacrum: [130°–160°]
Lumbar: [100°–150°]
Dorsal: [45°–100°]
Cervical: [0°–45°]

Right Thumb (Lung Lymph Runaround)
Axis Angle: 80°
Teeth, Odonton, 3rd Quadrant: [0°–180°]
31, 32, 33, 34, 35, 36, 37, 38
Teeth, Odonton, 2nd Quadrant [180°–360°]
28, 27, 26, 25, 24, 23, 22, 21
Nose, Ethmoid: [345°–15°]
Ear, TMJ, Tonsil: [165°–195°]
Sinus, Maxillary Sinus, Frontal Sinus: [270°]

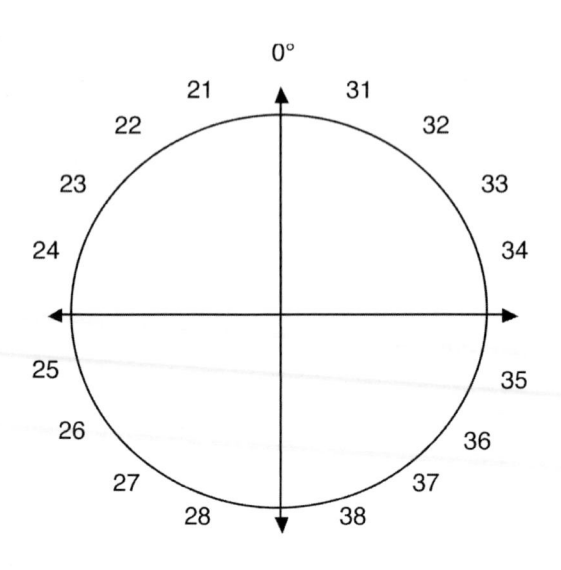

Left Thumb (Lung-Lymph-Runaround)

Axis Angle:	80°
Teeth, Odonton, 1st Quadrant:	[0°–180°]
31, 32, 33, 34, 35, 36, 37, 38	
Teeth, Odonton, 4th Quadrant:	[180°–360°]
28, 27, 26, 25, 24, 23, 22, 21	
Nose, Ethmoid:	[345°–15°]
Ear, TMJ, Tonsil:	[165°–195°]
Sinus, Maxillary Sinus, Frontal Sinus:	[90°]

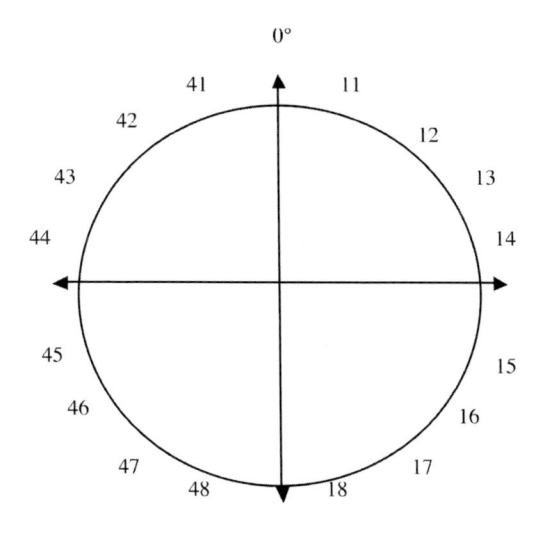

Energy-Organ Relationships of the Toes

Since in a photon-diagnosis as in the ETD by Mandel the radiations at the toe tips are recorded as well, more energy-organ relationships have emerged empirically from the plurality of thousands of shots in practice, whose topography is listed below.

Left Big Toe (Liver-Spleen-Pancreas Meridian)
Latent Diabetes: [0°–90°]
Pancreas – Ferments and Enzymes: [90°–180°]
Head Zone, Pancreas-Liver: [150°–210°]
Spleen: [340°–20°]

Right Big Toe (Liver-Spleen-Pancreas Meridian)
Genetic Diabetes: [270°–360°]
Liver - Parenchyma: [0°–180°]
Head Zone, Pancreas-Liver: [150°–210°]
Pancreas-Ferments and Enzymes: [180°–270°]

Left 2nd Toe (Stomach-Joint-Connective Tissue-Degeneration)
Joint Degeneration: [0°–180°]
Acid Behavior – Mucose: [180°–360°]

Right 2nd Toe (Stomach-Joint- / Connective Tissue-Degeneration)
Stomach – Pylorus: [0°–180°]
Joint–Degeneration: [180°-360°]

Left 3rd Toe (Skin- / Connective Tissue-Degeneration)
Neuroglia: [0°–360°]
Skin-Degeneration: [180°–360°]
Scar: [160°–200°]

Right 3rd Toe (Skin-/Connective Tissue-Degeneration)
Body-Connective-Tissue – Basic System: [0°–360°]
Skin Degeneration: [0°–180°]
Scar: [160°–200°]

Left 4th Toe (Bile – Oily Degeneration)
Small Bile Duct System: [0°–360°]
Fat Metabolism: [0°–360°]

Right 4th Toe (Bile – Oily Degeneration)
Gall-Bladder: [0°–360°]
Fat-Metabolism: [0°–360°]

Left 5th Toe (Bladder – Kidney Meridian)
Urogenital: [0°–360°]
Kidney: [0°–180°]
Bladder: [180°–360°]
Head Zone – Urogenital: [160°–200°]

Right 5th Toe (Bladder – Kidney Meridian)
Urogenital: [0°–360°]
Bladder: [0°–180°]
Kidney: [180°–360°]
Head Zone – Urogenital: [160°–200°]

For these empirically derived energy-organ relationships that have been known since the early 1980s from ETD, there is currently no physically based model of explanation. From the Traditional Chinese Medicine (TCM) and other traditions is known that several areas of the body of man correspond to each other in some mysterious way. In the ear acupuncture the anatomical details of the ear match certain body areas, as for example the earlobe matches the head and the outer edge of the auricle matches the course of the human spine. In the reflexology massage anatomical details of the palms or soles are assigned to specific organs within the body.

From conventional medicine it is known that the human embryo develops from the fertilized ovum of the mother. Initially, there are three germ layers created, from which then later in the development are formed the different types of tissue, such as skin tissue, digestive tract and connective tissue; of which in the further course the specialized organs of the human body are built. So perhaps, the key to understanding these anatomical correspondences lies in the cotyledon development of the human embryo.

Since the topography of the luminous phenomena occurring during the Kirlian effect ultimately depends on the distribution of positive and negative electrical charge carriers at the skin surface, ie, the terminal points, a physically based model must take this into account. Here, more basic research is needed in the coming years. An approach in solid state physics in the description of the charge distributions in a micro circuit may offer one starting point for a physical model. I was occupied with this during my PhD thesis at Siemens AG in Munich-Neuperlach. In such a design model the different conductive layers of a microchip can be represented by an equivalent circuit image to which its electrical behavior may be simulated in a relatively simple way.

So, it would be an interesting research project to model the human body in the form of an electronic equivalent circuit image.

CHAPTER VII

Results of Initial Case Studies

For some time now, a new generation of digital photon diagnostic systems have been tested in the medical practice. The research group of Dr. Jörg Kastner with Dr. Michael Jack in the practice of General & Holistic Medicine in Munich has participated. Meanwhile, the pictures of fingers and toes of many patients have been taken before and after treatments. Some representative cases of this ongoing study will be published in advance for the first time in this chapter. These first results underline the performance and usefulness of this technology in the complimentary medical diagnosis in the medical practice, especially in the documentation of the electrical charge in the cells of a patient before and after a treatment.

All images were taken in constant device parameters. These include the settings of the digital camera, the amplitude and frequency of the electrical voltage and the position of the fingers and toes at the recording electrode as well as the exposure time and the humidity of the exposure field. Therefore, the intensity and quality of the recorded radiations can be compared with each other.

Subsequently, Dr. Kastner has reported, based on three representative case studies, how he used a modern photon diagnosis system in the context of the patient history and to document the course of therapy in his practice.

Case study 1

The patient reported in the first interview that she was under examination stress. She could not sleep well any more and that she had the feeling, her head would be blocked. Furthermore, she was afraid of not getting all the facts in her head and not being able to pass the test. She said that she felt powerless and driven by an inner restlessness. After food intake she had abdominal pain and diarrhoea more often now as well as increased back pain.

Anamnesis:

Female patient, 26 years old, medical student, a lot of stress, exam anxiety, inner restlessness, lack of self-esteem, fear of failure, sleep disorders, feeling unable to free the head, recurring infections, recidivous throat inflammations, weak immune system, partial loss of appetite, gastrointestinal ailments, recurrent diarrhoea, recurrent pain in the upper and lower back, cold hands, cold feet, general weakness. Otherwise no abnormal medical history. Physical testing without medical evidence. The first recording of the hands with the photon diagnosis system, shown in Figure 18a and b, results in

the following picture (in this case we will analyze the individual points in this book only roughly to get an overview and general understanding. The photon diagnostic procedures and the following recordings allow still for much more detailed statements): Mixture of endocrine and toxic radiation quality, with significant gaps and lack of radiation in each finger area. This means: Blocked and lacking energy in many body areas.

Little finger, heart – small – intestine meridian:
Left little finger blocked in the range of t he aggressive zone and the heart energy.
Ring finger, triple heater – psyche – meridian:
Interpretation: This so-called upper area especially represents heart and psyche.
Here significant gaps are showing up on the photograph: Little finger left: aggressive zone, thyorgenic zone as well as a cloudy brightening little finger right: for the area of the vessels.
Interpretation: Significant signs of stress with uncoordinated energy in the head region; another symptom: Insomnia.
Ring finger left: Striking that there is a very incomplete and lacking radiation in the field of the pancreas, adrenal gland and up to the triple heater.

Ring finger right: Also highly patchy and a lack of radiation range for adrenal cortex, thymus, pituitary, triple heater. **Interpretation:** Longer-lasting strong fatigue with stress on the organs that secrete hormones and the stress hormone; incipient threat of a strong depletion condition. This impression is reinforced by the absence of radiation on the right middle finger for the area of the kidney.

In addition, striking that *small finger right:* lack of radiation of the jejunum and duodenum.

Index finger right: limited radiation and lack of radiation in the colon area (large intestine).

Interpretation: This gives rise to not only gastro-intestinal discomfort and indigestion, but may also cause a bad gut flora, which is responsible for a good immune system. Based on this bad energy picture, we have to assume a weakened or highly disturbed immune system.

This then shows up in the right thumb: Missing and disordered radiation in the ear, tonsils and lymphatic pharyngeal ring.

The patient indicated recurrent infection-susceptibility. Furthermore, the photon-image shows in the area of the *index finger left*: gaps in the radiation field for the back, cervical spine, lumbar and sacral.

The patient reported increasing back pain.

Patient from Case Study 1 before Treatment

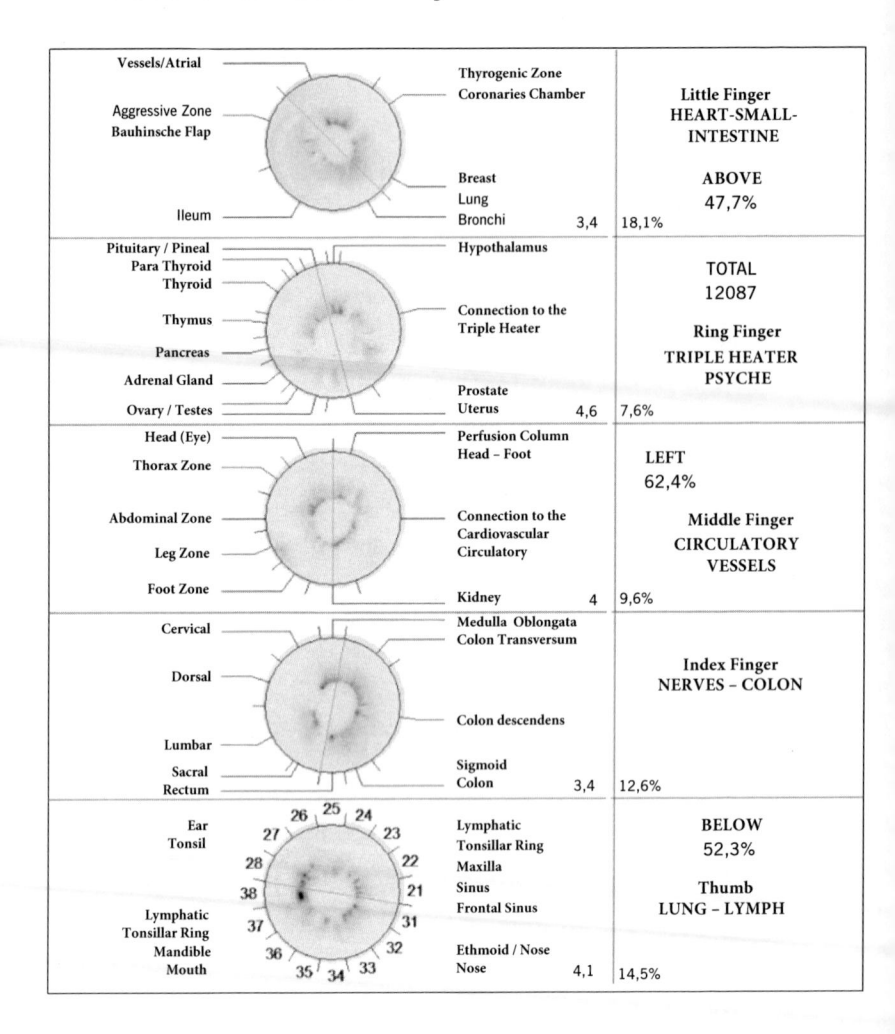

Vessels/Atrial Aggressive Zone Bauhinsche Flap Ileum	Thyrogenic Zone Coronaries Chamber Breast Lung Bronchi	3,4	18,1%	Little Finger HEART-SMALL-INTESTINE ABOVE 47,7%
Pituitary / Pineal Para Thyroid Thyroid Thymus Pancreas Adrenal Gland Ovary / Testes	Hypothalamus Connection to the Triple Heater Prostate Uterus	4,6	7,6%	TOTAL 12087 Ring Finger TRIPLE HEATER PSYCHE
Head (Eye) Thorax Zone Abdominal Zone Leg Zone Foot Zone	Perfusion Column Head – Foot Connection to the Cardiovascular Circulatory Kidney	4	9,6%	LEFT 62,4% Middle Finger CIRCULATORY VESSELS
Cervical Dorsal Lumbar Sacral Rectum	Medulla Oblongata Colon Transversum Colon descendens Sigmoid Colon	3,4	12,6%	Index Finger NERVES – COLON
Ear Tonsil Lymphatic Tonsillar Ring Mandible Mouth	Lymphatic Tonsillar Ring Maxilla Sinus Frontal Sinus Ethmoid / Nose Nose	4,1	14,5%	BELOW 52,3% Thumb LUNG – LYMPH

Fig. 18a: Left Hand

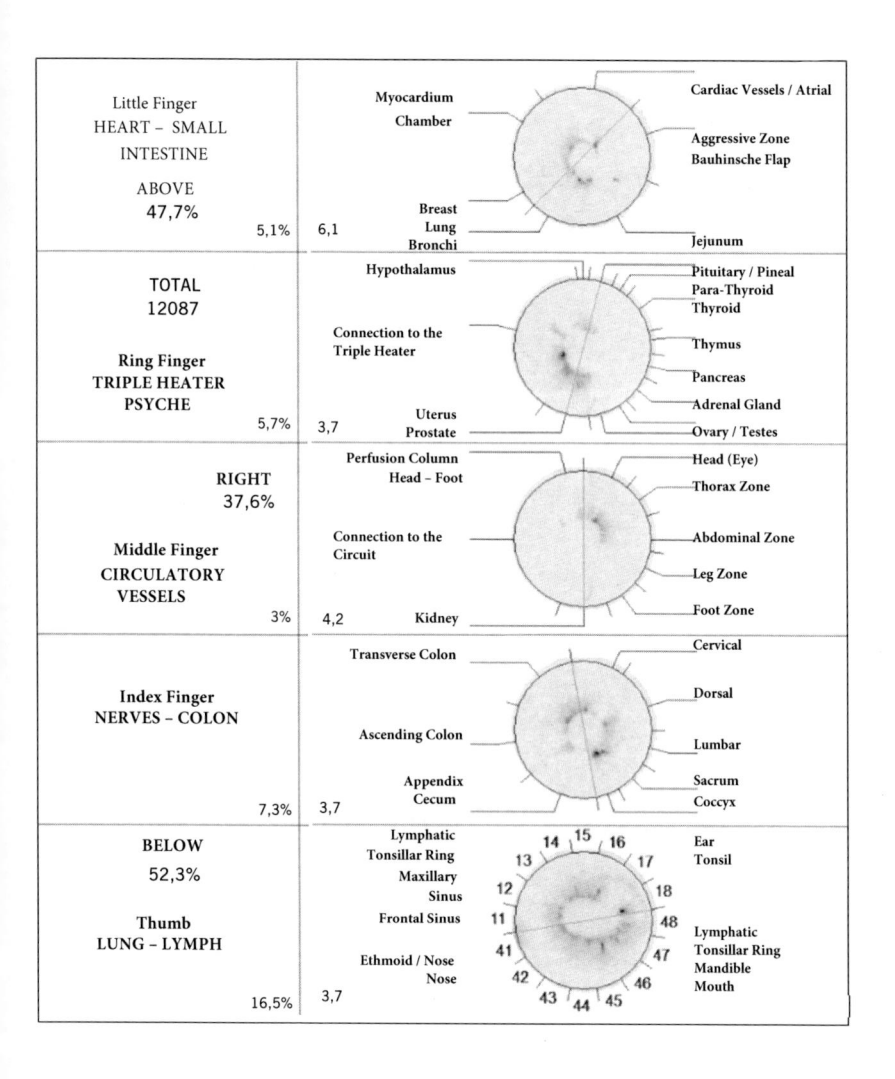

Little Finger HEART – SMALL INTESTINE ABOVE 47,7% 5,1%	6,1	Myocardium Chamber Breast Lung Bronchi		Cardiac Vessels / Atrial Aggressive Zone Bauhinsche Flap Jejunum
TOTAL 12087 Ring Finger TRIPLE HEATER PSYCHE 5,7%	3,7	Hypothalamus Connection to the Triple Heater Uterus Prostate		Pituitary / Pineal Para-Thyroid Thyroid Thymus Pancreas Adrenal Gland Ovary / Testes
RIGHT 37,6% Middle Finger CIRCULATORY VESSELS 3%	4,2	Perfusion Column Head – Foot Connection to the Circuit Kidney		Head (Eye) Thorax Zone Abdominal Zone Leg Zone Foot Zone
Index Finger NERVES – COLON 7,3%	3,7	Transverse Colon Ascending Colon Appendix Cecum		Cervical Dorsal Lumbar Sacrum Coccyx
BELOW 52,3% Thumb LUNG – LYMPH 16,5%	3,7	Lymphatic Tonsillar Ring Maxillary Sinus Frontal Sinus Ethmoid / Nose Nose	14 15 16 13 17 12 18 11 48 41 47 42 46 43 44 45	Ear Tonsil Lymphatic Tonsillar Ring Mandible Mouth

Fig. 18 b: Right Hand

Patient from Case Study 1 after Treatment

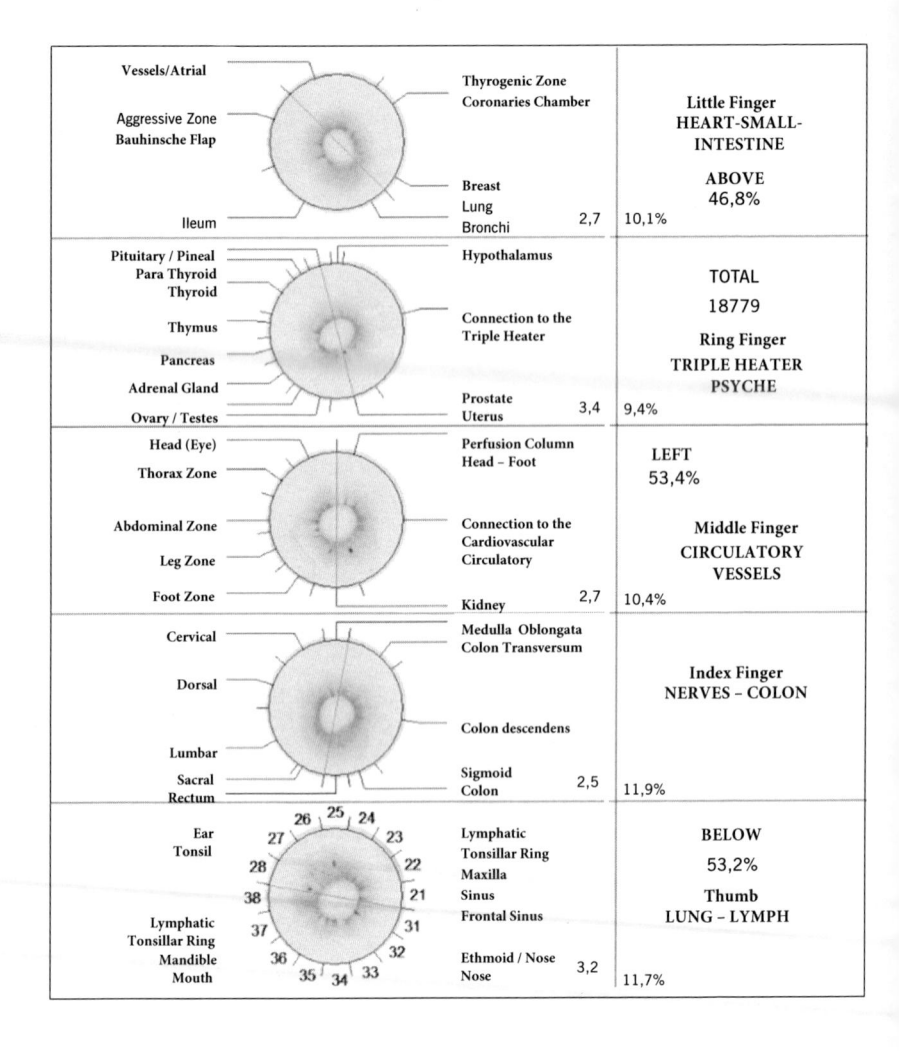

Vessels/Atrial Aggressive Zone Bauhinsche Flap Ileum		Thyrogenic Zone Coronaries Chamber Breast Lung Bronchi 2,7	10,1%	Little Finger HEART-SMALL- INTESTINE ABOVE 46,8%
Pituitary / Pineal Para Thyroid Thyroid Thymus Pancreas Adrenal Gland Ovary / Testes		Hypothalamus Connection to the Triple Heater Prostate Uterus 3,4	9,4%	TOTAL 18779 Ring Finger TRIPLE HEATER PSYCHE
Head (Eye) Thorax Zone Abdominal Zone Leg Zone Foot Zone		Perfusion Column Head – Foot Connection to the Cardiovascular Circulatory Kidney 2,7	10,4%	LEFT 53,4% Middle Finger CIRCULATORY VESSELS
Cervical Dorsal Lumbar Sacral Rectum		Medulla Oblongata Colon Transversum Colon descendens Sigmoid Colon 2,5	11,9%	Index Finger NERVES – COLON
Ear Tonsil Lymphatic Tonsillar Ring Mandible Mouth		Lymphatic Tonsillar Ring Maxilla Sinus Frontal Sinus Ethmoid / Nose Nose 3,2	11,7%	BELOW 53,2% Thumb LUNG – LYMPH

Fig. 19 a: Left Hand

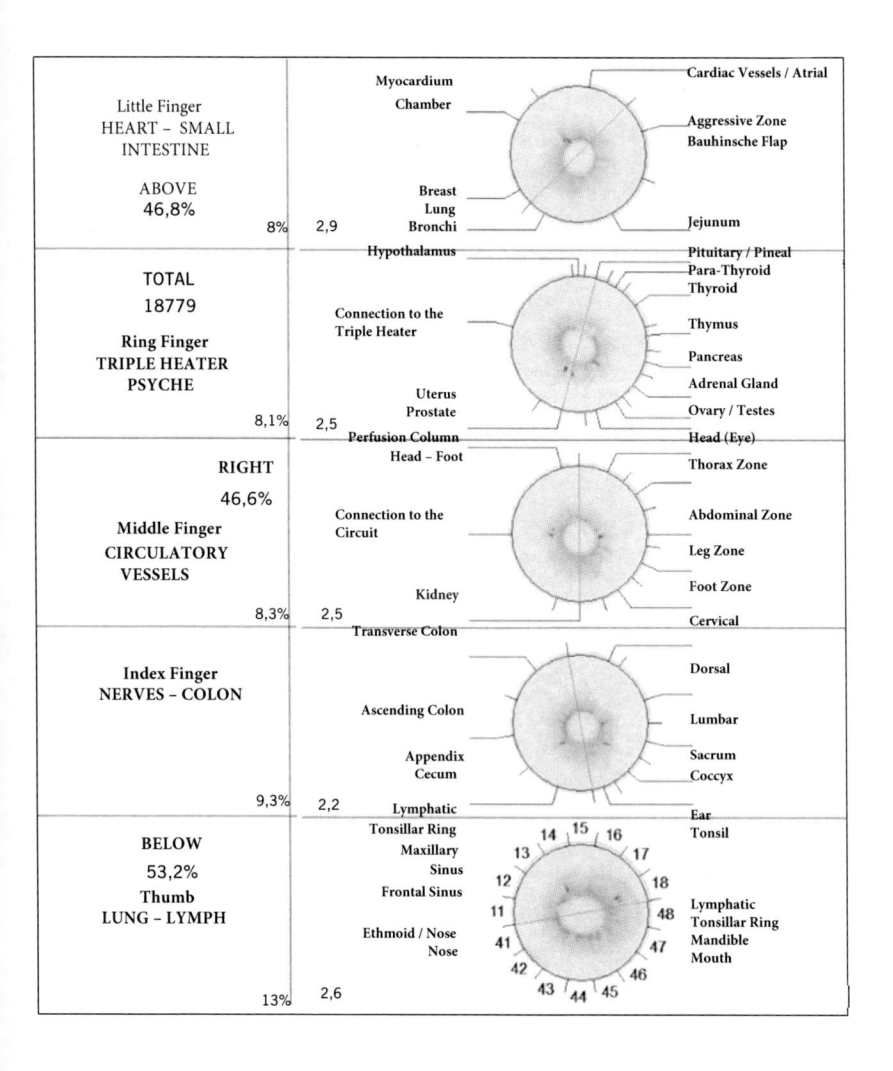

Little Finger HEART – SMALL INTESTINE ABOVE 46,8% 8%	2,9	Myocardium Chamber Breast Lung Bronchi		Cardiac Vessels / Atrial Aggressive Zone Bauhinsche Flap Jejunum
TOTAL 18779 Ring Finger TRIPLE HEATER PSYCHE 8,1%	2,5	Hypothalamus Connection to the Triple Heater Uterus Prostate Perfusion Column		Pituitary / Pineal Para-Thyroid Thyroid Thymus Pancreas Adrenal Gland Ovary / Testes Head (Eye)
RIGHT 46,6% Middle Finger CIRCULATORY VESSELS 8,3%	2,5	Head – Foot Connection to the Circuit Kidney Transverse Colon		Thorax Zone Abdominal Zone Leg Zone Foot Zone Cervical
Index Finger NERVES – COLON 9,3%	2,2	Ascending Colon Appendix Cecum Lymphatic		Dorsal Lumbar Sacrum Coccyx Ear
BELOW 53,2% Thumb LUNG – LYMPH 13%	2,6	Tonsillar Ring Maxillary Sinus Frontal Sinus Ethmoid / Nose Nose	14 15 16 13 17 12 18 11 48 41 47 42 46 43 44 45	Tonsil Lymphatic Tonsillar Ring Mandible Mouth

Fig. 19 b: Right Hand

Summary interpretation:

The patient shows a markedly patchy and weak overall energetic picture in many areas. The anamnestic data of the patient is confirmed in the photon-diagnosis. This also shows us that the patient threatens to drift into an even greater and very serious state of exhaustion, because important endocrine organs (endocrine glands) are blocked, or have too little energy (kidney, adrenal gland, pituitary gland).

Due to the weakened intestinal energy, the patient is also thretened by an alarming weakening of the immune system.

Furthermore, a blockage and a deficiency in the field of heart energy/psyche is shown here and is something that needs to be treated.

Of course, some conclusions result from the mere medical history, but with the aid of the photon-diagnosis, much more detail is shown, which, if properly linked, allows a much more efficient therapy.

Figures 19a and b show the second recording of the fingers with the photon-diagnosis-system (right after the 30-minute treatment, about 35 minutes after the first shot). The treatments were primarily in the field of naturopathic medicine: Acupuncture, Interference Field Therapy, Homeopathy as well as a series of Vitamin-Support-Infusions.

Interpretation: On all fingers the radiation coronae are clear to see and they are closed almost everywhere. Where there was previously no visible radiation, emissions are now showing.

You can see very clearly the difference after just thirty minutes the energy of the patient is flowing very well and almost all organ systems are again supplied sufficiently or even well.

In this regard, one can encourage the exhausted patient, as she essentially still has sufficient reserves and also responded very well to the treatments. This is definitely not always the case. Similarly, one can also work with more difficult reaction types, only this time more specifically and ask them from the beginning, to have a little more patience and they can also be motivated if they use the changes in the pictures according to the treatments so that they follow the therapy consistently and do something for themselves.

Patient 1 had therefore sufficient energy reserves, they just had to enable the flow again.

The ring finger right shows a cloudy deposit in the area of the pituitary gland, which still indicates a restless, unstable mental condition. For this reason the patient received a stabilizing homeopathic remedy for the psyche to use at home.

The further images of the patient showed an improving stabilization of her energetical state. The treatment methods could be used very selectively due to the detailed results of the photon diagnosis. In addition, the quantitative evaluation of the test results showed an increase of the measured emission from 12087 radiation units (before treatment) to 20097 radiation units (after treatment). This represents an increase of 66 percent.

Case Study 2

The patient came to the pracitce with acute allergic symptoms. She reported severe itching and swelling of the hands and forearms. Furthermore, she complained of itching around the eyes and the neck.

Anamnesis:

Patient, 36 years old, single mother, five-year-old son, employed, periods with a lot of stress, lately increasing amounts of mild allergic reactions with skin and eyes itching, but no shortness of breath. The allergic reaction may be due to food, strong sunlight and sudden temperature differences, coupled with stress. She was sleeping irregularly, often interrupted and not enough. Increasing loss of energy, recurrent infections, cold feet, overall the patient is very sensitive to cold.

Examination Findings:

Hands reddened, swollen, slightly red eyes, lungs are clear, no other major examination findings.

Patient 2 before Treatment:

The first picture of the hands taken with the photon-diagnosis-system (Figure 20a and b) results in the following image: Mix of endocrine and toxic radiation quality, overall weak emission.

Most important Conspicuousnesses:
Ring finger left: Adrenal gland, ovary and thymus.
Little finger right: Lung area, bronchi, jejunum.
Ring finger right, as well as on the *left ring finger:* ovary, adrenal cortex, thymus
Middle finger right: kidney

Interpretation:

Overall weak energy levels; specific deficits in the lungs and intestines (jejunum) which are important for the immune system and therefore the reason for allergic reactions: furthermore, there may be a food intolerance, which is also often a cause of allergic reactions, since there exists an energetic weakness in the digestive tract as well as in the hormone-producing organs – ovaries, thymus, adrenal cortex.

This case also raises a suspicion of a significant overall hormonal imbalance, which could also explain the fatigue.

The follow up treatment was carried out exclusively on the basis of naturopathy, such as acupuncture and homeopathy. After 30 minutes the control picture followed.

Patient from Case Study 2 before the treatment :

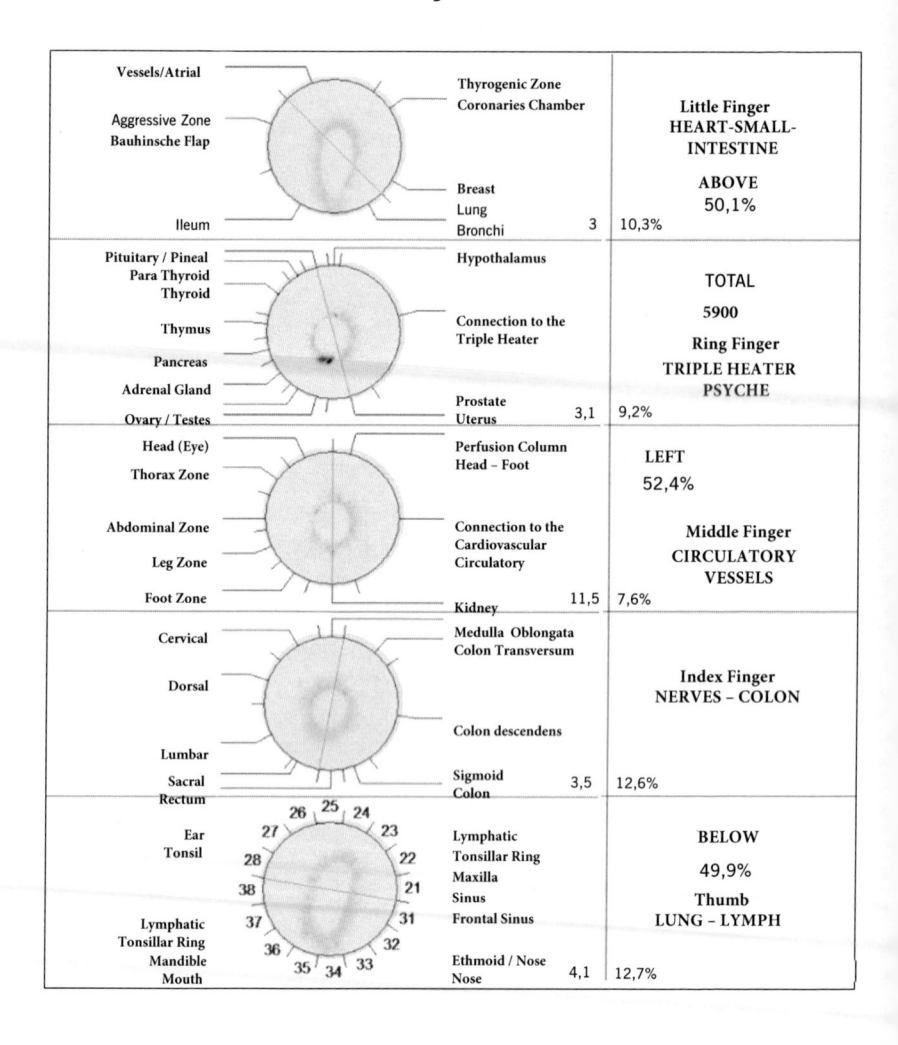

Vessels/Atrial Aggressive Zone Bauhinsche Flap Ileum		Thyrogenic Zone Coronaries Chamber Breast Lung Bronchi 3	Little Finger HEART-SMALL-INTESTINE ABOVE 50,1% 10,3%
Pituitary / Pineal Para Thyroid Thyroid Thymus Pancreas Adrenal Gland Ovary / Testes		Hypothalamus Connection to the Triple Heater Prostate Uterus 3,1	TOTAL 5900 Ring Finger TRIPLE HEATER PSYCHE 9,2%
Head (Eye) Thorax Zone Abdominal Zone Leg Zone Foot Zone		Perfusion Column Head – Foot Connection to the Cardiovascular Circulatory Kidney 11,5	LEFT 52,4% Middle Finger CIRCULATORY VESSELS 7,6%
Cervical Dorsal Lumbar Sacral Rectum		Medulla Oblongata Colon Transversum Colon descendens Sigmoid Colon 3,5	Index Finger NERVES – COLON 12,6%
Ear Tonsil Lymphatic Tonsillar Ring Mandible Mouth	26 25 24 27 23 28 22 38 21 37 31 36 32 35 34 33	Lymphatic Tonsillar Ring Maxilla Sinus Frontal Sinus Ethmoid / Nose Nose 4,1	BELOW 49,9% Thumb LUNG – LYMPH 12,7%

Fig. 20a: Left Hand

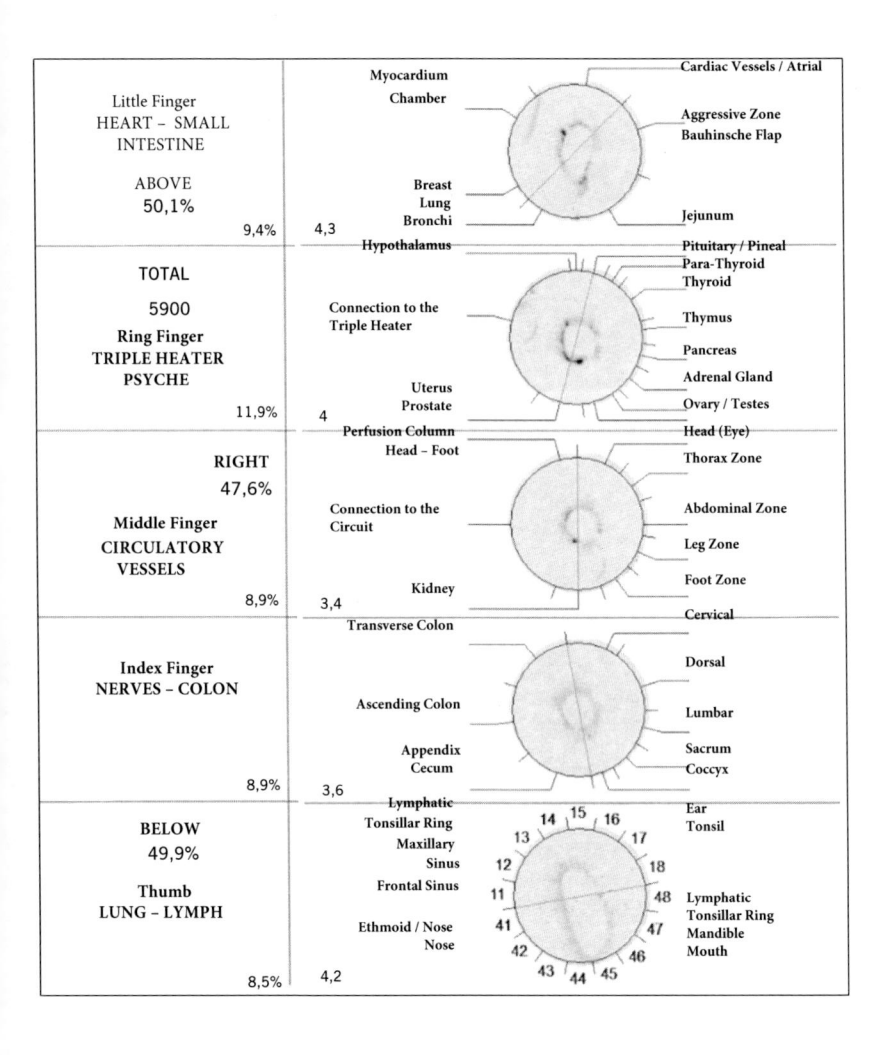

Little Finger HEART – SMALL INTESTINE ABOVE 50,1% 9,4%	4,3	Myocardium Chamber Breast Lung Bronchi	Cardiac Vessels / Atrial Aggressive Zone Bauhinsche Flap Jejunum
TOTAL 5900 Ring Finger TRIPLE HEATER PSYCHE 11,9%	4	Hypothalamus Connection to the Triple Heater Uterus Prostate	Pituitary / Pineal Para-Thyroid Thyroid Thymus Pancreas Adrenal Gland Ovary / Testes
RIGHT 47,6% Middle Finger CIRCULATORY VESSELS 8,9%	3,4	Perfusion Column Head – Foot Connection to the Circuit Kidney	Head (Eye) Thorax Zone Abdominal Zone Leg Zone Foot Zone
Index Finger NERVES – COLON 8,9%	3,6	Transverse Colon Ascending Colon Appendix Cecum	Cervical Dorsal Lumbar Sacrum Coccyx
BELOW 49,9% Thumb LUNG – LYMPH 8,5%	4,2	Lymphatic Tonsillar Ring Maxillary Sinus Frontal Sinus Ethmoid / Nose Nose	Ear Tonsil Lymphatic Tonsillar Ring Mandible Mouth

Fig. 20b: Right Hand

Patient from Case Study 2 after the Treatment:

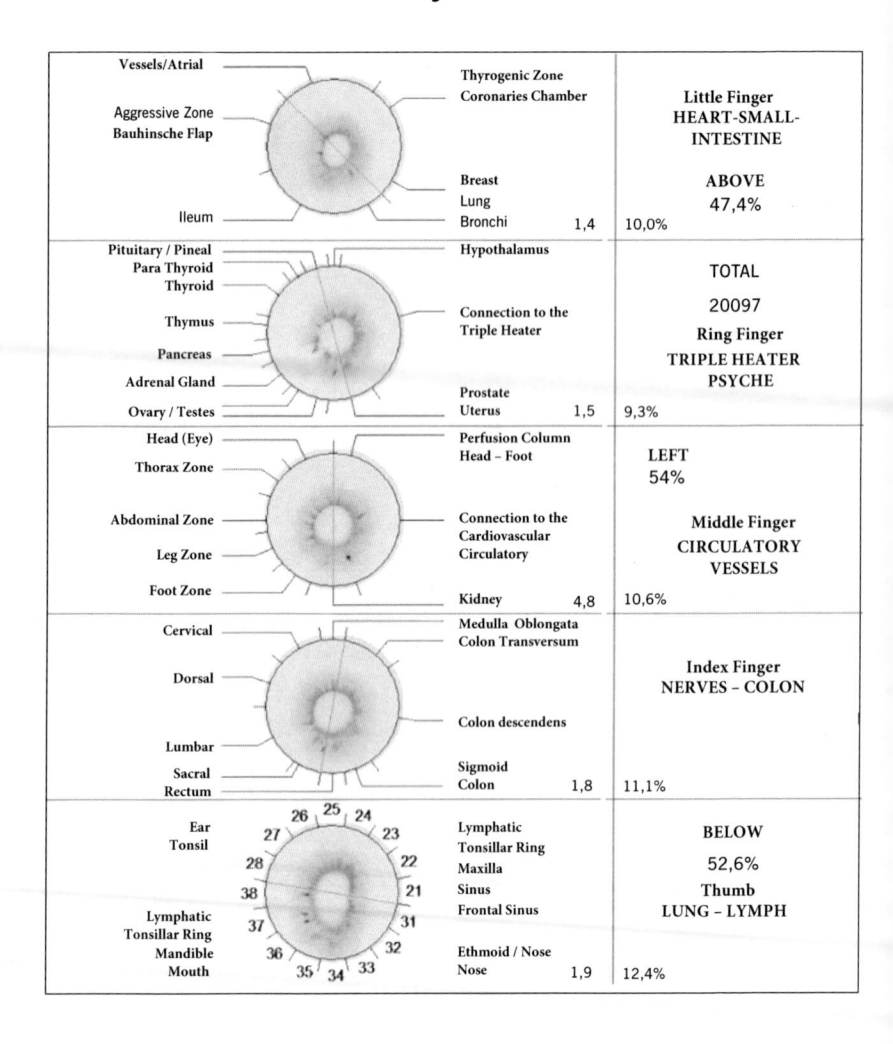

Vessels/Atrial Aggressive Zone Bauhinsche Flap Ileum	Thyrogenic Zone Coronaries Chamber Breast Lung Bronchi 1,4	**Little Finger** **HEART-SMALL-** **INTESTINE** ABOVE 47,4% 10,0%
Pituitary / Pineal Para Thyroid Thyroid Thymus Pancreas Adrenal Gland Ovary / Testes	Hypothalamus Connection to the Triple Heater Prostate Uterus 1,5	TOTAL 20097 **Ring Finger** **TRIPLE HEATER** **PSYCHE** 9,3%
Head (Eye) Thorax Zone Abdominal Zone Leg Zone Foot Zone	Perfusion Column Head – Foot Connection to the Cardiovascular Circulatory Kidney 4,8	LEFT 54% **Middle Finger** **CIRCULATORY** **VESSELS** 10,6%
Cervical Dorsal Lumbar Sacral Rectum	Medulla Oblongata Colon Transversum Colon descendens Sigmoid Colon 1,8	**Index Finger** **NERVES – COLON** 11,1%
Ear Tonsil Lymphatic Tonsillar Ring Mandible Mouth	Lymphatic Tonsillar Ring Maxilla Sinus Frontal Sinus Ethmoid / Nose Nose 1,9	BELOW 52,6% **Thumb** **LUNG – LYMPH** 12,4%

Fig. 21a: Left Hand

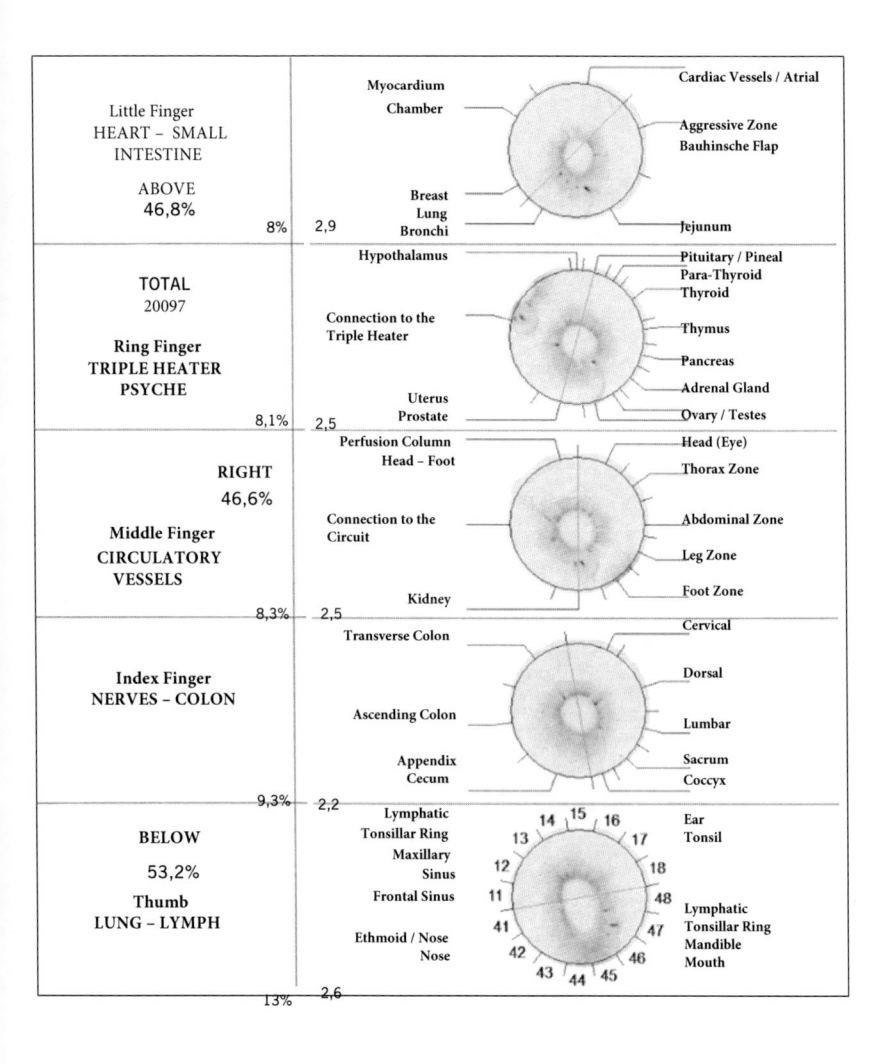

Little Finger HEART – SMALL INTESTINE ABOVE 46,8% 8%	2,9	Myocardium Chamber Breast Lung Bronchi	Cardiac Vessels / Atrial Aggressive Zone Bauhinsche Flap Jejunum
TOTAL 20097 Ring Finger TRIPLE HEATER PSYCHE 8,1%	2,5	Hypothalamus Connection to the Triple Heater Uterus Prostate	Pituitary / Pineal Para-Thyroid Thyroid Thymus Pancreas Adrenal Gland Ovary / Testes
RIGHT 46,6% Middle Finger CIRCULATORY VESSELS 8,3%	2,5	Perfusion Column Head – Foot Connection to the Circuit Kidney	Head (Eye) Thorax Zone Abdominal Zone Leg Zone Foot Zone
Index Finger NERVES – COLON 9,3%	2,2	Transverse Colon Ascending Colon Appendix Cecum	Cervical Dorsal Lumbar Sacrum Coccyx
BELOW 53,2% Thumb LUNG – LYMPH 13%	2,6	Lymphatic Tonsillar Ring Maxillary Sinus Frontal Sinus Ethmoid / Nose Nose	Ear Tonsil Lymphatic Tonsillar Ring Mandible Mouth

Fig. 21b: Right Hand

The recording with the photon-diagnosis-system after the treatment (Figure 21a and b) shows:

Significantly improved radiation intensity, but further gaps.

Little finger right: Area psyche (heart myocardium) and bronchi jejunum; equally striking *Ring finger right:* still with gaps in the field of the hormon producing organs.

Summary Interpretation:

The patient has reacted quite well to the treatment, feels better and the itching has ceased.

However, there remain further deficits in the area of the Little finger right: Psyche long term charged, unstable, signs of exhaustion; psyche should be absolutely stable. Furthermore, there are significant weaknesses in the intestinal region with indications of a possibly impaired intestinal flora (this should be ascertained by conventional medicine). All of this is important as a whole for the immune system and in the case of allergic tendencies; here, further therapy and support are urgently needed.

There are also deficits in the area of the hormon-producing organs, here further conventional medical investigation is nessary, for example, through specific laboratory diagnosis.

Note: The photon diagnosis gave here valuable information, among others, to check the patients intestinal functions, intestinal flora, as well as the hormonal system. The subsequently performed conventional medical diagnostic procedures confirmed the indications of the photon diagnosis.

Case Study 3

The patient reported about an increasing number of digestiv problems in the past three months, with abdominal bloating and distension, usually half an hour after eating. In addition he felt tired and exhausted afterwards. Overall, he felt an ever decreasing quality of life in the last six months, although he was sleeping enough and had no particular additional stress.

Anamnesis:
46-year-old patient in a good, relatively balanced overall condition, no specific test results. There were also no abnormal blood values found in the last six months, despite extensive laboratory diagnostics.

The fist recording with the photon diagnosis system (Figure 22a and b) resulted in:
Mostly a mixture of endocrine and toxic radiation quality with significant breaks in the coronae in the following areas:

Little finger left: Entire range unstable
Ring finger left: Gap in the digestive tract and pancreas
Middle finger left: Abdominal zone
Index finger left: Rectum and colon area
Thumb left: Lymphatic area, tonsils
Little finger right: Also generally instable, significant gap in the jejunum, gastrointestinal.

Patient from Case Study 3 before Treatment:

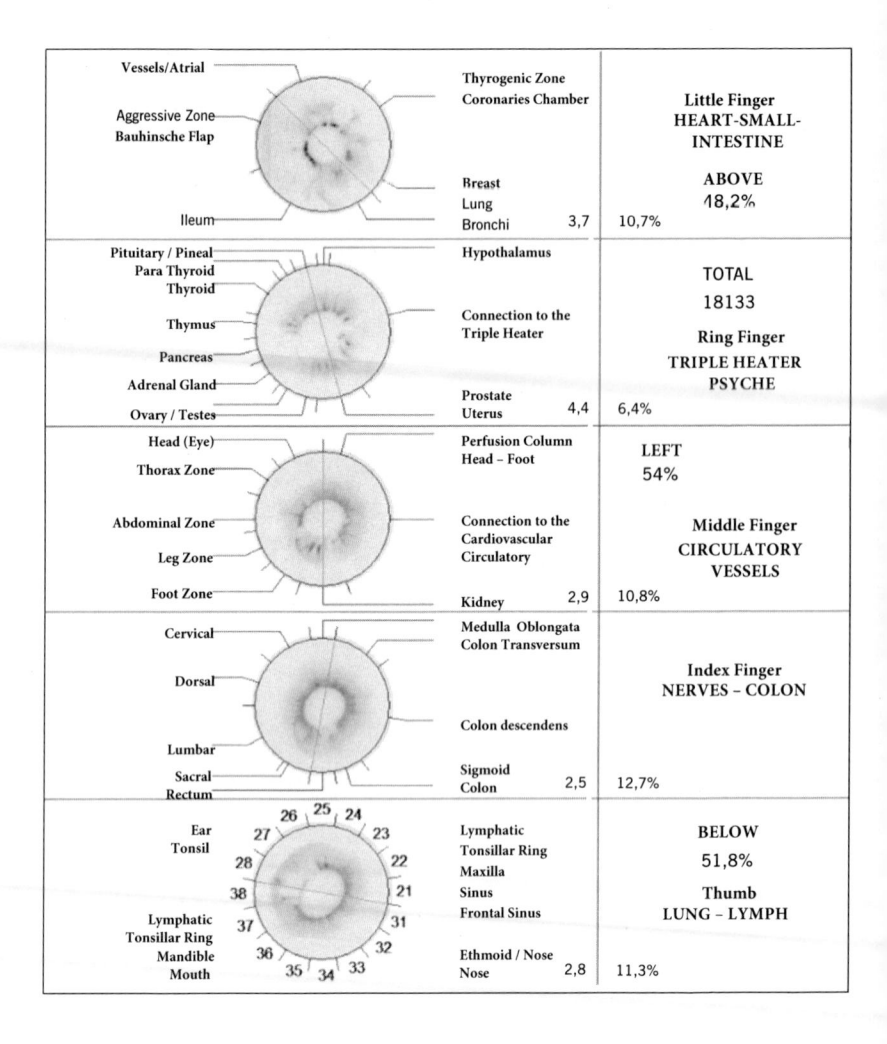

Diagram labels (left)	Diagram labels (center)	Value	%	Finger / Region
Vessels/Atrial; Aggressive Zone; Bauhinsche Flap; Ileum	Thyrogenic Zone; Coronaries Chamber; Breast; Lung; Bronchi	3,7	10,7%	Little Finger — HEART-SMALL-INTESTINE — ABOVE 18,2%
Pituitary / Pineal; Para Thyroid; Thyroid; Thymus; Pancreas; Adrenal Gland; Ovary / Testes	Hypothalamus; Connection to the Triple Heater; Prostate; Uterus	4,4	6,4%	TOTAL 18133 — Ring Finger — TRIPLE HEATER PSYCHE
Head (Eye); Thorax Zone; Abdominal Zone; Leg Zone; Foot Zone	Perfusion Column Head – Foot; Connection to the Cardiovascular Circulatory; Kidney	2,9	10,8%	LEFT 54% — Middle Finger — CIRCULATORY VESSELS
Cervical; Dorsal; Lumbar; Sacral; Rectum	Medulla Oblongata; Colon Transversum; Colon descendens; Sigmoid Colon	2,5	12,7%	Index Finger — NERVES – COLON
Ear; Tonsil; Lymphatic Tonsillar Ring; Mandible; Mouth (21–38)	Lymphatic Tonsillar Ring; Maxilla; Sinus; Frontal Sinus; Ethmoid / Nose; Nose	2,8	11,3%	BELOW 51,8% — Thumb — LUNG – LYMPH

Fig. 22a: Left Hand

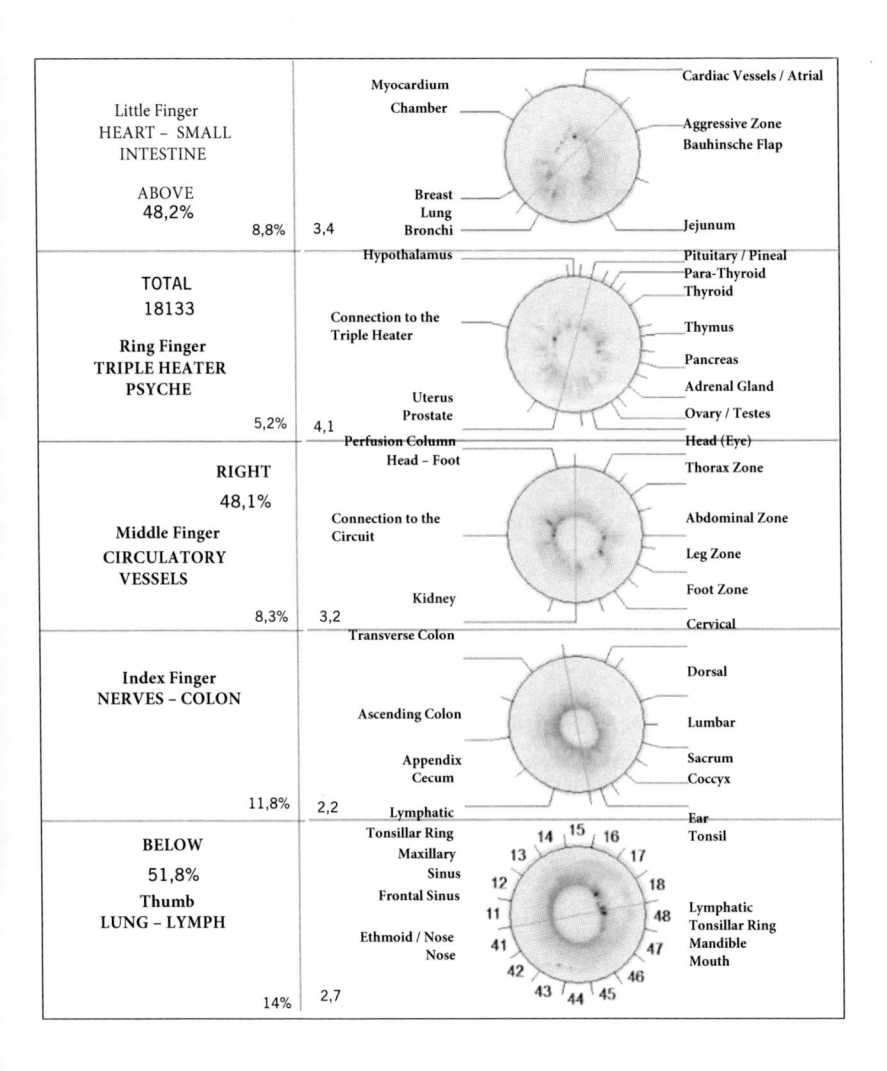

Little Finger HEART – SMALL INTESTINE ABOVE 48,2% 8,8%	3,4	Myocardium Chamber Breast Lung Bronchi	Cardiac Vessels / Atrial Aggressive Zone Bauhinsche Flap Jejunum
TOTAL 18133 Ring Finger TRIPLE HEATER PSYCHE 5,2%	4,1	Hypothalamus Connection to the Triple Heater Uterus Prostate	Pituitary / Pineal Para-Thyroid Thyroid Thymus Pancreas Adrenal Gland Ovary / Testes
RIGHT 48,1% Middle Finger CIRCULATORY VESSELS 8,3%	3,2	Perfusion Column Head – Foot Connection to the Circuit Kidney	Head (Eye) Thorax Zone Abdominal Zone Leg Zone Foot Zone Cervical
Index Finger NERVES – COLON 11,8%	2,2	Transverse Colon Ascending Colon Appendix Cecum Lymphatic	Dorsal Lumbar Sacrum Coccyx Ear
BELOW 51,8% Thumb LUNG – LYMPH 14%	2,7	Tonsillar Ring Maxillary Sinus Frontal Sinus Ethmoid / Nose Nose	Tonsil Lymphatic Tonsillar Ring Mandible Mouth

Fig. 22b: Right Hand

Patient from Case Study 3 after Treatment :

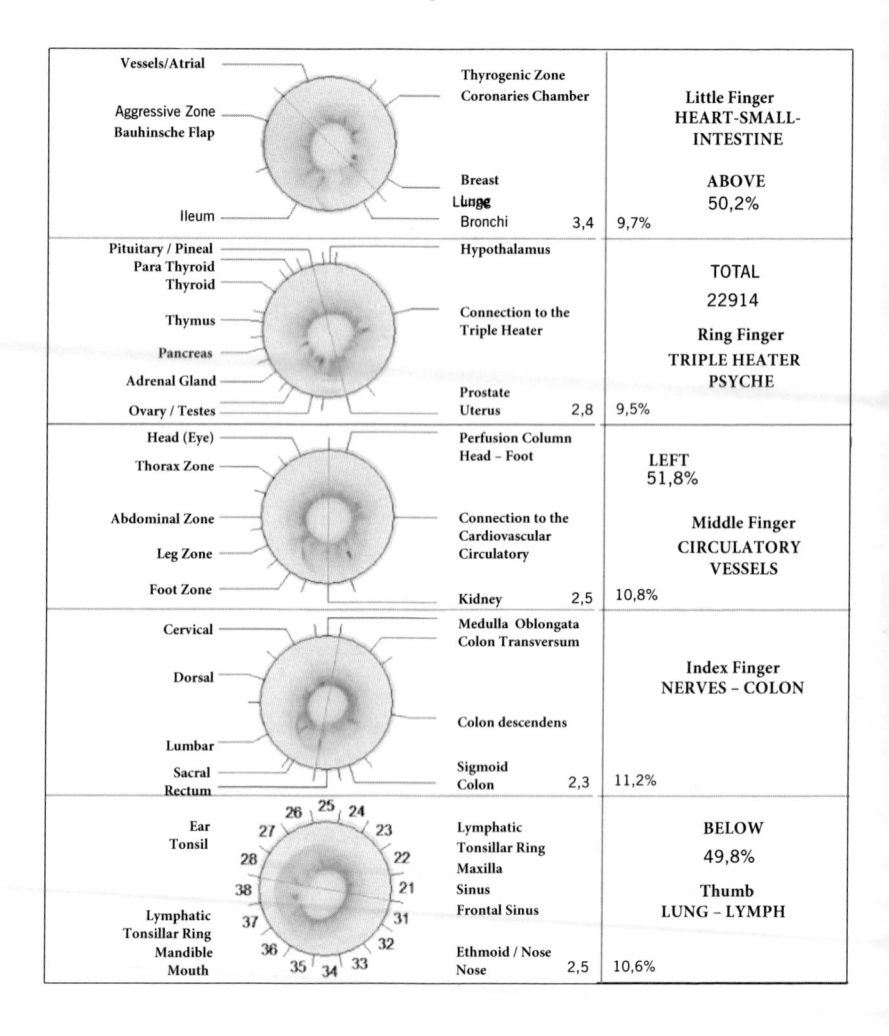

Vessels/Atrial Aggressive Zone Bauhinsche Flap Ileum	Thyrogenic Zone Coronaries Chamber Breast Lunge Bronchi 3,4	9,7%	Little Finger HEART-SMALL- INTESTINE ABOVE 50,2%
Pituitary / Pineal Para Thyroid Thyroid Thymus Pancreas Adrenal Gland Ovary / Testes	Hypothalamus Connection to the Triple Heater Prostate Uterus 2,8	9,5%	TOTAL 22914 Ring Finger TRIPLE HEATER PSYCHE
Head (Eye) Thorax Zone Abdominal Zone Leg Zone Foot Zone	Perfusion Column Head – Foot Connection to the Cardiovascular Circulatory Kidney 2,5	10,8%	LEFT 51,8% Middle Finger CIRCULATORY VESSELS
Cervical Dorsal Lumbar Sacral Rectum	Medulla Oblongata Colon Transversum Colon descendens Sigmoid Colon 2,3	11,2%	Index Finger NERVES – COLON
Ear Tonsil Lymphatic Tonsillar Ring Mandible Mouth	Lymphatic Tonsillar Ring Maxilla Sinus Frontal Sinus Ethmoid / Nose Nose 2,5	10,6%	BELOW 49,8% Thumb LUNG – LYMPH

Fig. 23 a: Left Hand

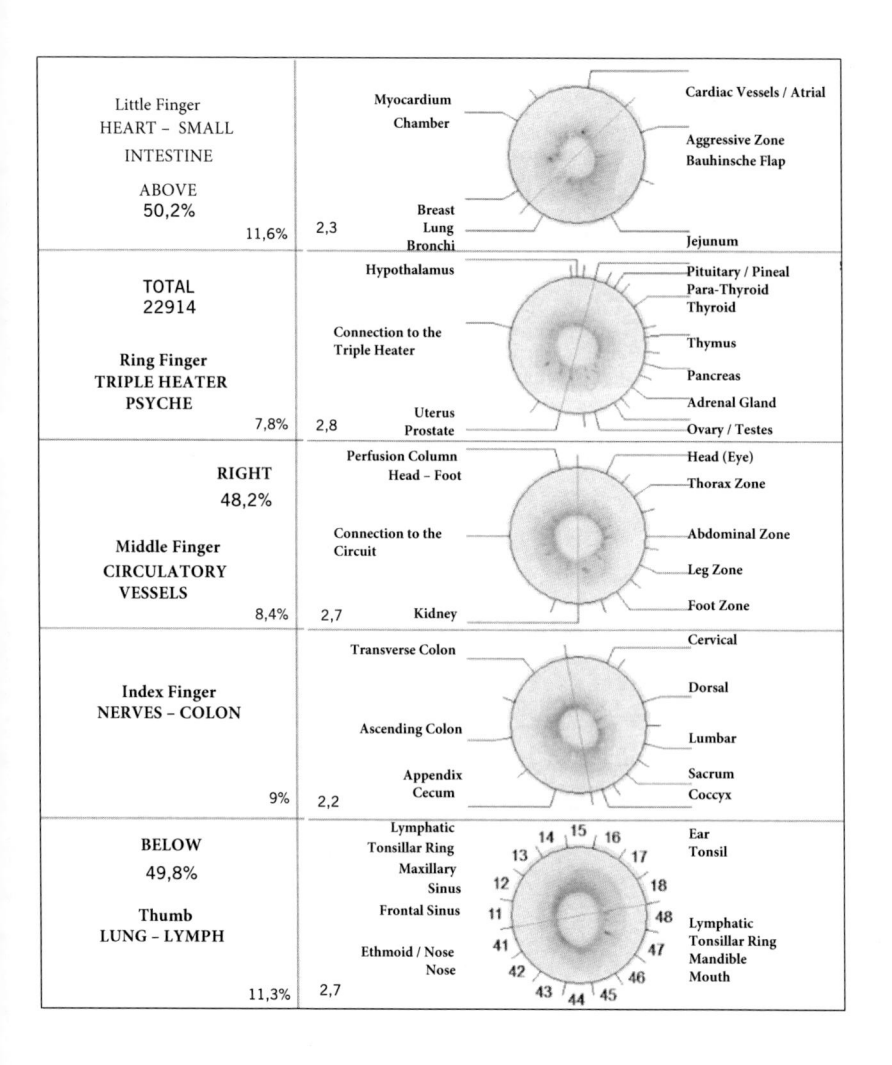

Little Finger HEART – SMALL INTESTINE ABOVE 50,2% 11,6%	2,3	Myocardium Chamber — Cardiac Vessels / Atrial — Aggressive Zone — Bauhinsche Flap Breast Lung Bronchi — Jejunum
TOTAL 22914 Ring Finger TRIPLE HEATER PSYCHE 7,8%	2,8	Hypothalamus — Pituitary / Pineal — Para-Thyroid — Thyroid Connection to the Triple Heater — Thymus — Pancreas — Adrenal Gland Uterus Prostate — Ovary / Testes
RIGHT 48,2% Middle Finger CIRCULATORY VESSELS 8,4%	2,7	Perfusion Column Head – Foot — Head (Eye) — Thorax Zone Connection to the Circuit — Abdominal Zone — Leg Zone Kidney — Foot Zone
Index Finger NERVES – COLON 9%	2,2	Transverse Colon — Cervical — Dorsal Ascending Colon — Lumbar Appendix — Sacrum Cecum — Coccyx
BELOW 49,8% Thumb LUNG – LYMPH 11,3%	2,7	Lymphatic Tonsillar Ring — Ear Maxillary — Tonsil Sinus Frontal Sinus — Lymphatic — Tonsillar Ring Ethmoid / Nose — Mandible Nose — Mouth

Fig. 23 b: Right Hand

111

Ring finger right: Patchy corona, especially in digestive organs and pancreas.

Interpretation:

It is striking that in the whole areas that represent the digestive organs, especially the pancreas and the abdominal zone, significant deficits are apparent.

Of course this coincides with the survey and the questionnaire of the patient. This can be well demonstrated pictorially and directly with the photon-diagnosis and be interpreted in the sense of an organ weakness or of an energy deficit.

Again, the first treatments administered were acupuncture and homeopathy. The following picture was taken also after a 30 minute treatment.

The second recording with the photon-diagnosis-system shows (Figure 23a and b): Overall a good response to the therapy with predominently closed, circular beams coronae with improved radiation.

Significantly improved emission and energy activity in the digestive organs and the pancreas after the treatment. There are still shortcomings in the little finger left and right in the zone of the ileum (intestine) and lung.

Summarized Interpretation:

The patient responded well to the treatment and, particularly in the digestive organs, the energy deficit has been reduced after the first treatment.

The patient felt more relaxed directly after the treatment but also "more energetic".

An energy deficit remains visible in the area of the intestine and lung/bronchi, which would need to be strengthened in further treatments.

General Note:

It is of course very interesting to see how these individual images change in the course of a few days or weeks, either in a positive or a negative sense. From these results extremly valuable insights on the energy or the health processes of the individual patients can be obtained. It should be noted that for a complete photon-diagnosis the foot must also be recorded, this has been deliberately omitted here.

The book should, at this point, only gives a small impression, a look through the keyhole so to speak, indicating what is potentially possible with the photon-diagnosis. At this point, it should be emphasized that we intend to carry out extensive studies to verify this method for the diagnostic and therapeutic range.

To anyone familiar with the photon picture interpretation let it be said again in short, that we only wanted to make a rough outline for the first general understanding in the evaluation of the images.

Summary:

From the first recordings of electrical discharges of living objects, which became known as Kirlian photographs, to the development of modern digital diagnostic procedures, which are used in complementary medicine, there was a vast development and several decades long. I am reporting in this book about the stages of the evolution and about my personal contribution to the Kirlian research.

For me it is clear that our vitality and mental state as well as our spirituality have become objectifiable with this procedure. From this we can gain medically relevant statements for the patients, which is impressively demonstrated through the results of the measurements made from a general medical practice which are published here in the case studies for the first time.

For this, a scientific consideration of the radiation phenomena occuring during the Kirlian effect as well as a technical advancement of the recording procedure were required.

I am convinced that the digital method of photon diagnosis which has evolved from the Kirlian photography has a great future in a medical practice, because it is perfectly suited for the documentation and visualization of the energetic effects of therapeutic applications.

Acknowledgements

I would like to thank the research group led by Dr. Jörg Kastner with Dr. Michael Jack in the practice of general and holistic medicine for their kindness and support in the presentation of contemporary examples from the medical practice.

Special thanks to Anna, Peter, Nick and Elliott, who helped me to translate the book into readable English.

For more information and seminars on the topic, please contact:

www.drmichaelkoenig.de
www.photonen-diagnose.de

List of Literature

Bergmann-Schäfer, Gobrecht, *Lehrbuch der Experimentalphysik, Band II, Elektrizität und Magnetismus,* Walter de Gruyter, Berlin, New York 1971

Popp, Fritz Albert, *Biologie des Lichts, Grundlagen der ultraschwachen Zellstrahlung,* Parey, Berlin, Hamburg, 1984

Büttrich, Sebastian und Gottschalk, Niels, *Kirlianfotografie,* Projektwerkstatt Physik, TU Berlin 1989/1990

Mandel, Peter, *Energetische Terminalpunkt-Diagnose,* Synthesis, Essen, 1983

Elektor, Zeitschrift für Elektronik, *Kirlianfotografie,* Mai 1977, S. 22–25

Strzempa-Depré (Michael König), *Verfahren zur Bestimmung der Verteilung und gegenseitigen Beeinflussung von positiven und negativen elektrischen Ladungen unter Ausnutzung des Kirlianeffekts.* Patentschrift DE 37 07 338 C2, Deutsches Patentamt, 1987

Strzempa-Depré (Michael König), *Transientes Latch-Up-Designmodell,* Dissertation, Universität Kassel 1986

Strzempa-Depré (Michael König), *Die Physik der Erleuchtung,* Goldmann, München, 1989

König, Michael, *Das Urwort – Die Physik Gottes,* Scorpio, München, 2010

König, Michael, *Der kleine Quantentempel,* Scorpio, München, 2011

König, Michael, *Photonen-Diagnostik als neueste Hightech-Weiterentwicklung des Kirlianverfahrens,* Comed, Fachmagazin für Complementär-Medizin, Kulmbach, Juli 2013